The Unknown Technology in Homer

HISTORY OF MECHANISM AND MACHINE SCIENCE

Volume 9

Series Editor

MARCO CECCARELLI

Aims and Scope of the Series

This book series aims to establish a well defined forum for Monographs and Proceedings on the History of Mechanism and Machine Science (MMS). The series publishes works that give an overview of the historical developments, from the earliest times up to and including the recent past, of MMS in all its technical aspects.

This technical approach is an essential characteristic of the series. By discussing technical details and formulations and even reformulating those in terms of modern formalisms the possibility is created not only to track the historical technical developments but also to use past experiences in technical teaching and research today. In order to do so, the emphasis must be on technical aspects rather than a purely historical focus, although the latter has its place too.

Furthermore, the series will consider the republication of out-of-print older works with English translation and comments.

The book series is intended to collect technical views on historical developments of the broad field of MMS in a unique frame that can be seen in its totality as an Encyclopaedia of the History of MMS but with the additional purpose of archiving and teaching the History of MMS. Therefore the book series is intended not only for researchers of the History of Engineering but also for professionals and students who are interested in obtaining a clear perspective of the past for their future technical works. The books will be written in general by engineers but not only for engineers.

Prospective authors and editors can contact the series editor, Professor M. Ceccarelli, about future publications within the series at:

LARM: Laboratory of Robotics and Mechatronics
DiMSAT – University of Cassino
Via Di Biasio 43, 03043 Cassino (Fr)
Italy
E-mail: ceccarelli@unicas.it

For other titles published in this series, go to
www.springer.com/series/7481

S.A. Paipetis

The Unknown Technology in Homer

Springer

S.A. Paipetis
Department of Mechanical Engineering
and Aeronautics
University of Patras
Patras 26500, Greece

From the original Greek "The Unknown Technology in Homer", Esoptron Publications, Athens, Greece, 2005.

Every effort has been made to contact the copyright holders of the articles and figures which have been reproduced from other sources. Anyone who has not been properly credited is requested to contact the publishers, so that due acknowledgements may be made in subsequent editions.

ISSN 1875-3442 e-ISSN 1875-3426
ISBN 978-90-481-2513-5 e-ISBN 978-90-481-2514-2
DOI 10.1007/978-90-481-2514-2
Springer Dordrecht Heidelberg London New York

Library of Congress Control Number: 2010926584

Printed on acid-free paper

Springer is part of Springer Science+Business Media (www.springer.com)

Contents

Part 2 Principles of Natural Science

Part 3 Automation and Artificial Intelligence

Part 4 Defensive Weapons in the Epics

Preface

Using such terms as science and technology, which have been relatively recently adopted, to write about situations and events that occurred 2,500 years ago, may be a paradox. The Homeric Epics, the Iliad and the Odyssey, refer to the Mycenean Era, a civilisation that flourished from the 16th to 12th century BCE. The seeming paradox ceases to be one when modern specialists, searching through the ancients texts, discover knowledge and applications so advanced, that can be termed as scientific or technological in the modern sense of the words.

The present book is based on extensive research performed by the author and his associates at the University of Patras, along with the presentations of other researchers at two international symposia, which he organized in Ancient Olympia.[1]

It consists of five parts, of which Part I is introductory, including such chapters as Homer and Homeric Epics, Troy and the mythological causes of the War, Achilles and his wrath, the siege and fall of Troy, Odysseus' long way home, the Trojan war and the cultural tradition, scientific knowledge in the Homeric Epics and finally an account on science and technology.

Part II includes three chapters on applications of principles of natural science, including chariot racing and the laws of curvilinear motion, creep in wood and hydrodynamics of vortices and the gravitational sling.

Part III consists of three chapters on automation and artificial intelligence, namely, on the forge of Hephaestus, the robots of Hephaestus and the Phaeacian ships and the UAVs.

[1] "Extraordinary Machines and Structures" (2001) and "Science and Technology in Homeric Epics" (2006).

Part IV deals with defensive weapons in the epics; its four chapters covering structural materials and analytical processess, the shield of Achilles, the shield of Ajax and other defensive weaponry.

Part V, in three chapters, deals with such specific subjects as the Trojan Horse, Mycenaean building and the admirable effect on physical health of reciting the epics with the proper Homeric meter.

Finally, in an appendix, *The Forge*, a literal-symbolic approach to the famous shield of Achilles is given, presenting the great ideas hidden in the construction and use of this mythical, magnificent masterpiece.

The results of this research oppose views of old, that, in the Homeric Epics hardly any significant elements of knowledge exist, and whatever is described as miraculous is poetic conception only. On some occasions, this may be true, but even then the need for specific applications of advanced knowledge is demonstrated, which is a powerful catalyst for technological progress. In other cases though, accounts of astonishing achievements are given, along with sufficient technical information, allowing for a thorough analysis by means of modern scientific methods and processes, and a more or less accurate evaluation of the technological background involved.

Out of this work, two important conclusions concerning the Homeric Epics can be drawn: (a) the scientific and technological knowledge they include is the result of interdisciplinary research, e.g., not of philology and/or archaeology alone, but of all scientific fields, (b) the Homeric Epics should be studied on a continual basis, since, with advancing sciences, new knowledge is constantly revealed, which, at earlier stages of scientific development, could easily pass unobserved.

Finally, there is no doubt that the Homeric Epics have affected art decisively, not only in the ancient world, but also in later times, especially in the West. In fact, they have been a constant source of inspiration for artists of all ages, and this is indeed useful, especially with optical art, presenting "artists' conceptions" of important scenes therein described, especially those of technological interest. Works of great painters and sculptors of all ages, depicting technological achievements accounted for in the Epics, are presented along with the respective analytical investigations.

S.A. Paipetis

Part 1
Introduction

Chapter 1
Homer and the Homeric Epics

So wast thou blind! – but then the veil was rent,
For Jove uncurtained Heaven to let thee live,
And Neptune made for thee a spumy tent,
And Pan made sing for thee his forest-hive.
John Keats, To Homer

1.1 The Homeric Epics

The Homeric Epics, before anything else, have taught us the Greek language properly – to such a level of excellence that they constitute world masterpieces. They have been the Gospel of the Hellenic people, a Gospel rid of elements of magic, metaphysics or superstition. Out of them, the Greeks have been taught models of honour, proper conduct and correct language.

Furthermore, the Homeric Epics have taught history, the history of the Minoan era and the beginning of the Mycenaean era. Although, in fact, they reflect the latter, while approaching its end, in a paradoxical way, they constitute the forerunner of a new era, providing the emerging Greek people a robust foundation on which they erected a new culture based on correct behaviour, pride and dignity.

The Greek civilization of the Homeric Epics gives the impression of something not radically new, but rather as the revival of the Aegean civilization, which, temporarily, almost disappeared due to violent clashes at the time. But knowledge, just like life, never dies completely, and, sustained by people's oral tradition, may lead to new understandings.

As far as the two Epics are concerned, the uniqueness of the *Iliad* lies in its early appearance and beauty. It follows the line of similar epics of other peoples and expresses their desire to trace their roots and their need

S.A. Paipetis, *The Unknown Technology in Homer*, History of Mechanism and Machine Science 9,
DOI 10.1007/978-90-481-2514-2_1, © Springer Science + Business Media B.V. 2010

Figure 1.1 Rembrandt, 1653: Aristotle with a bust of Homer, Metropolitan Museum of Art, New York (reproduced by permission).

to perpetuate the memory of great events, praised by inspired anonymous poets.

The *Iliad*, the most ancient monument of European literature, has been classified as a miraculous work, exquisitely perfect and of great length. It consists of 15,693 verses against 12,110 of the *Odyssey*. As a comparison, the *Aeneid* consists of 9,985 verses, Dante's *Divine Comedy* of 14,233, Milton's *Paradise Lost* of 10,565, *Erotocritus,* the Cretan epic of the 17th century, of 11,400 15-syllable verses, and *Digenis Akritas*, the Byzantine epic of the 12th to 13th centuries, less than 5,000, and this holds for all other known epics of the more recent European literature. On the other hand, the extravagance of the eastern world has led to much longer epics. For example, the *Mahabharata* consists of 220,000 verses, and the *Ramayana*, contemporary to the Iliad, of about 48,000. However, epics of the West conform better with human size and life duration.

The Iliad, as an exquisite literary monument, appears to have accumulated cultural efforts of many generations during many centuries. It is work like those old architectural monuments that survive in the form of magnificent mediaeval cathedrals. And, in fact, this epic did not appear at the end or the apogee of the Hellenic civilization, but rather at its beginning: Homer is the herald or the forerunner of the culture of Greece, of Europe and of the Western world in general. A herald of such a grand stature that even today overshadows anybody else.

Odyssey, the second great epic, appears to have assumed its final form several decades after the Iliad. According to some scholars, there is a good chance that the author may be a person different than that of the Iliad, however, one cannot ignore the possibility to be written by the same author at an older age. Both have a lot in common from the point of view of vocabulary, grammar, rhetoric, and prosody, as well as some extraordinary common properties, such as simplicity of thought and formulation, which distinguishes them from the slowness and indolence of the epics of the East.

Between Iliad and Odyssey there are considerable differences, mainly of tone and atmosphere. The first is a polemic narrative: the clashes between persons and warriors are formidable. Accordingly, it contains exceptionally important technological elements. The inventions, the tricks, but also a deep knowledge of techniques, appear to approach the frontier of modern technology, and are found there in abundance.

On the contrary, Odyssey is governed by a completely different atmosphere. It is full of peace, human feelings, traveling, magic, imagination, moral teachings. And all this within a world full of conflicts, trials and struggle. In the same way that, in a more symbolic form, the story of Odysseus was interpreted by C. Cavafy.[1] The Odyssey is the first work of fiction in world's literature. It is clear that the two epics are separated by a long period of peace, during which the military technological achievements were utilized for peaceful purposes, leading to social and cultural development. Finally, the Odyssey reflects the inner disposition of a man, who, after many misfortunes glorious experiences

[1] Constantine P. Cavafy (1863–1933), a major non-conformist Greek poet of Alexandria (Egypt). His famous poem *Ithaki* (1911) is based on the voyage of Odysseus back to his home island. The idea is to enjoy the journey and learn from it, which is more important than arriving at the destination and that maturity of the soul is all one can ask for.

πολλῶν δ' ἀνθρώπων εἶδεν ἄστεα καὶ νόον ἔγνω

many cities did he visit, and many were the nations with whose manners and customs he was acquainted

is looking forward to its end, having acquired the peaceful wisdom that the voyage itself has donated him. Concerning the real duration of this intermediate period, of the symbolic 20 years of traveling, the following remark holds: Although both epics refer to the Bronze Age, this metal is mentioned in the Iliad fourteen times more than iron, while in the Odyssey four times only, which may denote one century between the completion of the two epics.

1.2 Homer

The questions of who that most eminent poet of antiquity actually was, what his particular characteristics were, when he lived and how different he was from other trobadours, remain unanswered. For over three centuries, specialists are dealing with the question, whether a person named Homer did indeed exist. Many claimed that "Homer" is only a collective name for a group of trobadours, the *Homeridae*, who, around 800 BCE, simply executed the contexture and the presentation of a circle of pre-existing oral epics. On the contrary, based on the study of the texts, others believe that only one person composed and presented the two major Homeric works.

What is certain, is that, one way or another, Homer did indeed exist, and this is confirmed by the admirable uniformity of the Iliad, the work of one single extraordinarily gifted person. There is a question though about the exact time that the Iliad was completed. The events described are believed to have taken place between 1280 and 1180 BCE, and, consequently, the poem was completed much later. Many of the techniques mentioned there, clearly existed prior to the Trojan War, however, the conviction prevails that the epic could never have been completed before the 10th or even the 9th century BCE.

Upon maturing of Greek culture, Homer, although unknown, acquired such a fame that no one doubted his existence. People imagined him as a blind old man, singing his own compositions, however, the visual quality of his works can hardly agree with this view, unless he was blinded at a later stage of his life. On the other hand, in the Cumaic dialect, Ὅμηρος (Homer) means "blind", while in the Ionian dialect the verb ὁμηρεύω (homērevo)

Figure 1.2 Left, a bust of Homer, right, archaic statuette of the 7th century BCE of a blind poet-singer, most probably Homer.

means "to guide" and, accordingly, it refers to the leader or the poet. Eventually, Homer was not thought of as a mortal. His work exists, but the man cannot be found.

1.3 The Homeric Tradition

The beginning of the Homeric tradition is lost in the depths of time. The Iliad and the Odyssey have been kept alive through the songs of the bards and the trobadours, who performed them on all festive occasions. In the mid-6th century (540 BCE), Xenophanes of Colophon states that 'Εξ ἀρχῆς καθ' Ὅμηρον μεμαθήκασι πάντες (from the beginning from Homer all have learnt), while, half a century later, at Pindar's time, as stated, some rhapsodists were called Ὁμηρίδες (Homeridae): 'Ομηρίδαι ῥαπτῶν ἐπέων ἀηδοί *(Homeridae singers of composed*[2] *epics)* – in reality spiritual sons of Homer and conservators of his tradition.

[2] Literally "stitched together".

Figure 1.3 Bust of Homer, Roman copy after a Helenistic original of the 2nd century BCE, Musei di Capitolini, Palazzo Nuovo, Rome (reproduced by permission).

The first regular Homeric text was located in Athens during the tyranny of Peisistratos, but after his death it was ignored or simply went astray. However, the tradition remained alive through its presentation among others, both at the annual *Panathenaia* and the five-yearly *Great Panathenaia*, with their musical contests. References to above text can be found in Herodotus, Plato and Xenophon. Two more Greek editions (διορθώσεις =

corrections) are mentioned, one by Antiochos of Claros (Ionia, end of Peloponnesian War) and the other by Aristotle in honor of Alexander the Great. It was only in the Hellenistic times that a scientific study of the text was initiated. The first "corrector" was Zenobius of Ephesus (3rd century BCE), first librarian of Alexandria Museum, considered to have completed the first edition of the Iliad and the Odyssey before 274 BCE. It also seems that he was the one who devised the separation of each of the epics into 24 books or *rhapsodies.*

Aristophanes of Byzantium and Aristarchus of Samothrace (2nd 1st century BCE), third and fourth librarians of Alexandria Museum respectively, improved the text substantially giving it its final form. However, "corrections" continued, and the history of Homeric teaching is characteristic of the very history of Hellenic education.

In its *Symposium (c, 5)*, Xenophon, speaking through the lips of a table-companion, states:

> ʽΟ πατὴρ ἐπιμελούμενος ὅπως ἀνήρ ἀγαθός γενοίμην,
> ἠνάγκασε μὲ πάντα τὰ ʽΟμήρου ἔπη μαθεῖν.
>
> My father, caring for me to become a good man,
> compelled me to learn all Homer's Epics.

However, Plato (*Republic*, 606e), although considering Homer the first and greatest tragedian, excludes him from his Republic. Despite this, Homer won the title of the *Master Teacher of the Hellenes*, a title respected even by the later anti-pagan prejudice of the Christians. In fact, Homer remains the Master Teacher of Humanity.[3]

1.4 The Development of Writing

Neither in the Iliad nor in the Odyssey is there a clear reference to writing, with one exemption (Iliad VI, 168–169):

> πέμπε δὲ μὶν Λυκιῆνδε, πόρεν δ' ὁ γὲ σήματα λυγρὰ
> γράψας ἐν πίναχι πτυκτῷ θυμοφθόρα πολλὰ
>
> so he sent him to Lycia with lying letters of introduction,
> written on a folded tablet, and containing much ill against the bearer.[4]

[3] George Sarton, *Ancient Science through the Golden Age of Greece*, Dover Publications, Mineola, New York, 1980.

[4] The English translations of the Homeric texts are those by Samuel Butler.

Of course it is not clear what the meaning of γράψας (= *writer*) is. The early meaning of γράφω (*to write*) was ξύνω (*to scratch*), which, much later, assumed the meaning of χαράσσω (*to engrave*), σχεδιάζω (*to draft*). It is also known that the word ἀναγιγνώσκω *(to read)*, initially meant *to know well* or *to recognize*, while, in the sense of "reading", it was first used by Pindar (6th–5th century BCE). In the same way, the Syrian word βιβλίο (*book*) was used by Herodotus in the sense of *piece of paper* or *letter*, while, in the modern sense, it was first used by Aristotle. It is a fact that writing did exist when the Iliad was completed, certainly in the area of the Aegean, being of Cretan origin.[5] However, it was only used for inscriptions, legislative or magical texts, registration of inventories, accounts and other short texts of technical nature. Writing as a means of communication was put to general use many centuries after its invention, and epic poetry was one of the last fields of its application. Besides, papyrus was in general use in Greece at about the end of the 7th century BCE.

Of course, the question remains whether it would be possible at all to develop science and technology, even at an elementary level, without writing, ensuring preservation, transmission and further development of existing knowledge. This phenomenon is not new since the mediaeval builders have left no written traces of the first-class knowledge they possessed, allowing them to erect magnificent buildings, such as cathedrals, castles, defence works, etc. Accordingly, one arrives at the conclusion that, even scientific and technological knowledge may be preserved and propagated through oral tradition, confined within the circles of private guilds, organized more or less as secret societies and possessing technological knowledge as well as a philosophical or moral dimension.

1.5 Bards and Rhapsodists

Poets and bards were the bearers of oral tradition, travelling from one place to another and from one court to another, exhibiting their creations with a purpose to entertain and educate their audience. The innate tendency of human beings towards rhythm, led them to present their works in metric form.

[5] The Mycenaean language is the oldest form of Greek, and was used in Mycenae and in Crete from the 16th to the 11th centuries BCE, before the Dorian invasion. It was preserved in inscriptions on tablets in Linear B script, which was born in Crete in the 14th century BCE, mainly on clay tablets of Mycenae and Knossos. This script was decoded in 1952 by Michael Ventris, and it was proved beyond doubt that it was an early form of the Greek language.

The metre, i.e., the recitation rhythm, besides its basic property, to provide the listener with a feeling of safety and peace,[6] it also operates as a "carrier wave", which is modulated by the trains of words which get adapted to it. In this way, not only is it easier to memorize, but also protects the text from corruption during oral transmission. The latter was the only way to preserve the racial and national traditions, at a time that writing and means of writing were not yet invented.

Peoples, at their infantile cultural stage, possessed many of the features of the infants of nowadays. They liked to listen to stories and were particularly attracted by the rhythm and the beauty of presentation. The element of surprise and the recognition of archetypal models in the depths of their own mind was a source of joy. The presentation was often assuming a ritual form, with repetition as its main characteristic, which was instinctive but also purposeful, since it reminded of or created words or phrases of wisdom, proverbs and dictums, guiding thought and behavior. Finally, the language was vivid and impressive and used powerful and elaborate forms of speech.

The development of the form of these poems depended on the particular bards. Others used to make creational interventions, to modify, to improve, make additions, while others were satisfied with the best possible presentation, according to their own view, to whatever they had been taught and knew how to present. Even in those cases, modifications were indispensable. Suffice to say that research on the performance of folk singers, proved that the same song was performed differently by different singers in the same times, and even different by the same singer at different times.

Homer as ἀοιδός (i.e., poet, sooth-sayer, prophet) is by far the best of all early bards. As a rhapsodist, it appears that a considerable part of his work consists of parts that he collected and put together (*stitched*), adding the product of his own inspiration, and eventually the magnificent final outcome. Occasional peculiarities, purposeless repetitions and imperfect transgressions appear to support this view. On the other hand, the ability of a bard, to memorize long poetry, a quality that modern man has nearly lost, was a substantial factor in preservation of the oral tradition.

---

[6] H.E. Huntley, *The Divine Proportion: A Study in Mathematical Beauty*, Dover Publications, New York, New York, 1970.

Chapter 2
Troy and the Mythological Causes of the War

Mycenaean Greeks of the 12th century BCE were contemporary to Troy, a city of Bronze Age (Figure 2.1), situated at the Aegean coast of Asia Minor. Both civilizations used to apply megalithic architecture.

Figure 2.1 Troy hillock (19th century engraving).

Heinrich Schliemann, a German businessman and archaeologist of the 20th century (Figure 2.2) conducted excavations both in Mycenae and in western Asia Minor, where he discovered the ruins of ancient Troy. Most probably, Troy was destroyed (also by fire) and rebuilt more than once (Figure 2.3).

Troy is one of the most famous sites of the western world, mainly due to its connection with Homer's Iliad. The city, at a first stage, seems to have been created during the 3rd millennium BCE. Around 3000 BCE, at its plane

S.A. Paipetis, *The Unknown Technology in Homer*, History of Mechanism and Machine Science 9, DOI 10.1007/978-90-481-2514-2_2, © Springer Science + Business Media B.V. 2010

Figure 2.2 Heinrich Schliemann.

Figure 2.3 Kerstiaen de Keuninck (Coninck), end 16th century: Fire of Troy, The State Hermitage Museum, St. Petersburg (Photograph © The State Hermitage Museum, reproduced by permission).

Figure 2.4 (a) The ruins of the walls of Troy and (b) their reconstruction.

level, a fortification wall along with single-brick wall buildings with stone foundations existed. The city fortification works were extended at the next stage in ca. 2700 BCE. Maintaining its cultural continuation, Bronze-Age Troy covered an area of about 2 hectares. Entrance was possible through two great gates protected by huge towers. Inside the walls, houses were much

Figure 2.5 Beads with fourfold spirals ca. 2500–2300 BCE from the area of Troy, Instambul Archaeology Museum (reproduced by permission).

Figure 2.6 Part of "Priam's treasures", discovered by Schliemann.

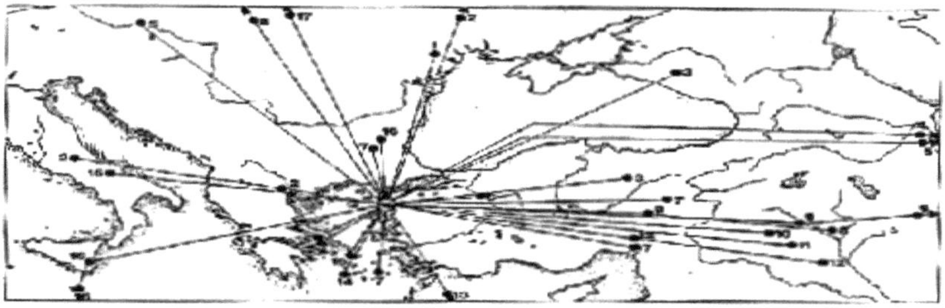

Figure 2.7 Distribution of goods from Troy to eastern Mediterranean.[1]

larger than in older times, built with sizeable stones. At this stage, Troy became legendary for its wealth (Figures 2.4–2.6).

This wealth grew thanks to an extended trade network, connecting Troy with other cities both in Anatolia and in the Cyclades Islands and the Greek mainland (Figure 2.7).

The mythological origin of Troy and the Trojan War are related to numerous ancient stories, legends and narratives, each one with many variations, often conflicting with each other. The Iliad covers a small part of them, as it deals with the *wrath* (μῆνις) of Achilles, the most important of the Greek heroes. Even the gods used to participate to the wars of the mortals, taking the part of one or another warrior. One should bear in mind that gods in ancient Greece did not perform miracles in the Judaeo-Christian sense, but only achievements, like those that heroes used to perform. On the other hand, the destiny of the mortals was not determined by the gods, but by the Fates, who had control even over the gods.

From all those stories, reference will be made only to those related with the purpose of the present book, e.g., connected to technological achievements or they created proper conditions for them.

2.1 The Mythological Causes of the Trojan War

Troy was erected by the gods Apollo and Poseidon, when a penalty inflicted upon them by Zeus for supposedly mutinous behavior, compelled them to work as mortals for Laomedon, Priam's father. They applied for assistance to Aeacus, mortal son of Zeus and Aegina, and grandfather of Achilles, since

[1] Korfmann, M. and Mannsperger, D. (1999). *Führer durch Troia: Troia, ein historischer Überblick und Rundweg*, Konrad Theiss Verlag, p. 78.

Fate had determined that Troy, erected by the hands of mortals, one day would fall in the hands of conquerors.

Indeed, soon after its erection, Troy was conquered by Heracles, Telamon, brother of Peleus and uncle of Achilles and also father of Ajax and Teucer, and by Peleus, son of Aeacus and father of Achilles. Their aim was to punish Laomedon, who did not deliver, as promised, his immortal horses to Heracles, who had saved the life of Hesione, his daughter. Telamon took Hesione, who was to give birth to Teucer.

Priam, king of Troy and son of Laomedon, was presented with a son from his wife Hecabe, who saw in a dream that she had given birth to a flaming torch. Cassandra, daughter of Priam and a soothsayer, predicted that the newborn son, Paris (or Alexander), would destroy the city, unless put to death upon birth. Paris was taken away from the city to be killed, but he was saved by some herdsmen and was brought up in the farms of Mount Ida. However, when grown up, he returned to Troy to take part to athletic games, he was recognized and returned to the royal family.

Peleus, father of Achilles, fell in love with Thetis, a sea-nymph (Figure 2.8), whom Zeus, father of gods, was personally interested in. However, Zeus was informed of an ancient prophecy stating that Thetis would bear a son who would become greater than his father. So he chose to concede to her

Figure 2.8 Peleus and Thetis, Museum of Fine Arts, San Francisco (reproduced by permission).

marrying Peleus, a mortal king. All gods were invited to the wedding, except Eris, goddess of strife and conflict. But not only did she arrive to the feast uninvited, she also brought a golden apple bearing an inscription "To the most beautiful" (Τῇ καλλίστῃ). The apple was claimed by Hera, spouse of Zeus, Aphrodite, daughter of Zeus, and Athena, also daughter of Zeus, who was asked, to choose the most beautiful. Zeus declined, so the burden of choice fell upon Paris, who was still a herdsman at Mount Ida. Each one of the goddesses promised Paris a precious reward, if he chose her. Hera offered him power, Athena polemic glory and wisdom, but Aphrodite offered him the most beautiful woman of the world: The apple was given to Aphrodite (Figures 2.9–2.11).

Figure 2.9 Jan Both (1615–1652) and Cornelis van Poelenburgh (1594/5–1667): Landscape with the judgement of Paris, National Gallery, London (reproduced by permission).

Helen was daughter of Zeus and Leda, who was approached by him in the form of a swan (Figures 2.12–2.15). Her beauty was famous all over the world. Tyndareus, husband of Leda, did not want Helen to marry a mortal, until all leaders of Greece vowed that they would together seek revenge for any insult to her. They did so and Helen then married Menelaus, king

Figure 2.10 Claude Lorraine (1600–1682): Judgement of Paris (1646), The National Gallerie of Art, Washington DC (reproduced by permission).

Figure 2.11 Rubens: Judgement of Paris, National Gallery, London (reproduced by permission).

Figure 2.12 Tintoretto: Leda and the swan, ca. 1555, Uffizi, Florence.

Figure 2.13 Leonardo: Leda and the Swan, (ca. 1505–1510) (replica by Cesare da Sesto), Earl of Pembroke, Wilton House, Salisbury, UK (reproduced by permission).

Figure 2.14 Correggio: Leda and the swan, ca. 1532, Gemäldegalerie, Berlin (reproduced by permission).

Figure 2.15 Gustave Moreau (1826 1898): Leda and the swan, Gustave Moreau Museum, Paris (reproduced by permission).

Figure 2.16 Assembly of gods before the Trojan War (360–350 BCE), vase discovered near Kerch (Bosporus Kingdom).

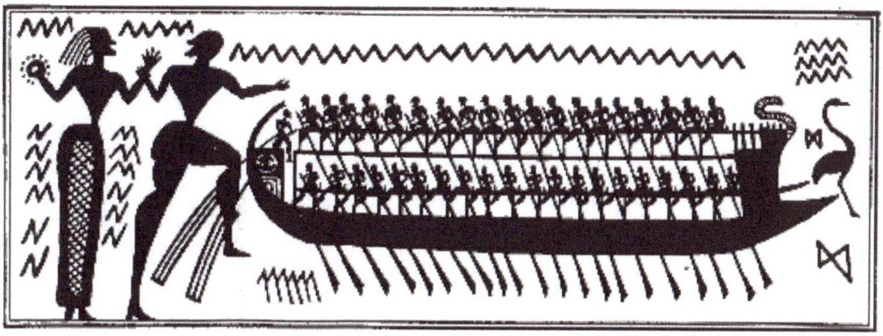

Figure 2.17 Flight of Helen with Paris: Theban vase, 8th century BCE.

Figure 2.18 Angelica Kauffman: Venus induces Helen to fall in love with Paris, The State Hermitage Museum, St. Petersburg (photograph © The State Hermitage Museum, reproduced by permission).

of Sparta. Her sister, Clytemnestra, daughter of Tyndareus, got married to Agamemnon, king of Argos and brother of Menelaus, and the most powerful leader in Greece.

Paris, now a royal prince, travelled to Sparta as ambassador of Troy, while Menelaus was away. Paris and Helen fell in love with each other most

Figure 2.19 Alecos Fassianos: Flight of Helen with Paris, artist's personal collection (reproduced by permission).

strongly, and flew together to Troy, carrying along considerable part of the city's treasure (Figures 2.17–2.20). Menelaus demanded that his wife and treasure be returned to him, and, upon Trojans' refusal, he invoked the vow of Tyndareus, compelling the Greeks to campaign against Troy. About one thousand ships took off from Aulis, Euboea, heading to Troy.

Figure 2.20 Paris and Helen, Attic Red-Figure Lekythos from terracotta, ca. 420–400 BCE. Attributed to painter Medias's circle, Athens. The J. Paul Getty Museum, Villa Collection, Malibu, California (reproduced by permission).

Chapter 3
Achilles and the Mēnis

Achilles was the son of Peleus and Thetis, a sea-nymph. Their marriage was permitted on the condition that their son, who was yet to be born, would die in war. Thetis, to protect the infant from death, bathed him in the waters of Styx, to make him invulnerable (Figures 3.1 and 3.2). Chiron, the Centaur, was Achilles's educator (Figures 3.3 and 3.4).

Figure 3.1 Donato Creti (1671–1749): Thetis dipping Achilles in the waters of Styx.

S.A. Paipetis, *The Unknown Technology in Homer*, History of Mechanism and Machine Science 9, DOI 10.1007/978-90-481-2514-2_3, © Springer Science + Business Media B.V. 2010

Figure 3.2 Rubens (draft for tapestry, 1760): Thetis dipping the infant Achilles into the river (i.e., of Styx), Boijmans van Beuningen Museum, Rotterdam (reproduced by permission).

Figure 3.3 Chiron and Achilles, red-painted amphora, ca. 480 BCE.

Figure 3.4 Pompeo Batoni (1708–1787), Chiron and Achilles.

Figure 3.5 Athenaean marble sarcophagus, 3rd century CE. Three of the walls consist of bas-relieves with scenes of the Iliad. In the picture, Achilles among Lycomedes' daughters, The State Hermitage Museum, St. Petersburg (Photograph © The State Hermitage Museum, reproduced by permission).

Figure 3.6 Pompeo Batoni, 1745, Achilles in the court of Lycomedes, Galleria degli Uffici, Florence.

Upon gathering of the Greek armies for the campaign against Troy, Achilles's parents decided to disguise him as a girl and hide him among the daughters of Lycomedes, king of Skyros island. There he met Dieidameia who gave him a son, Triptolemus. However, Calchas, the soothsayer, told Agamemnon and the other leaders that they would never conquer Troy without Achilles' support. Then Odysseus went to Skyros and managed to trick Achilles out of the girls' company, by putting a weapon in front of them. Achilles stretched out his arm to grasp it, revealing his true identity (Figures 3.5 and 3.6).

The Greek fleet could not depart from Aulis due to windlessness, until Agamemnon, the commander-in-chief, offered to sacrifice Iphigeneia, his daughter, to goddess Artemis. Another version tells us that at the last moment Iphigeneia was saved by Artemis and brought to Tauris to become a priestess of her.

Figure 3.7 Ajax and Achilles: Greek amphora, 510–490 BCE.

Figure 3.8 Achilles and Troilus: urn from Southern Italy (350–325 BCE).

Figure 3.9 Giovani Battista Tiepolo: The wrath of Achilles, 1757, Villa Valmarana, Vicenza (reproduced by permission).

Figure 3.10 Alecos Fassianos: The wrath of Achilles, artist's collection (reproduced by permission).

Figure 3.11 Giovani Battista Tiepolo: Eurebates and Talthyvius lead Bryseis to Agamemnon, 1757, Villa Valmarana, Vicenza (reproduced by permission).

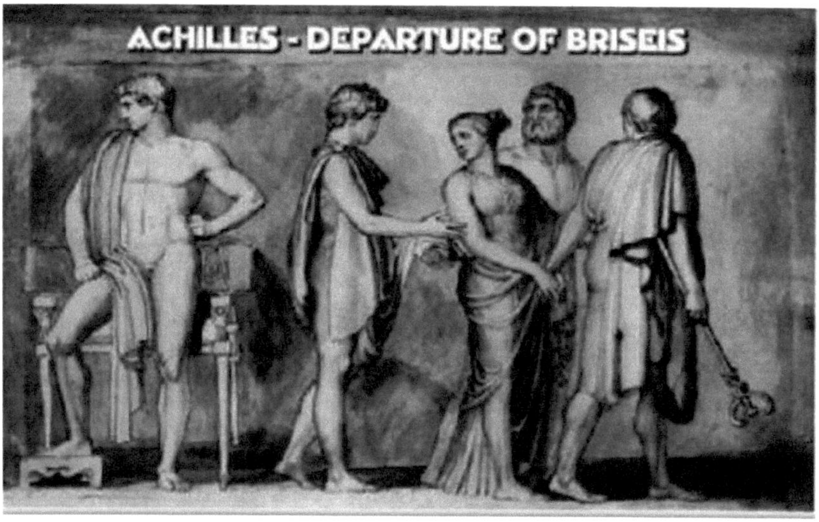

Figure 3.12 John Flaxman: Departure of Bryseis from Achilles (illustration in Pope's translation of the Iliad, 1805).

The Greek troops landed at the coast of Troy, which they besieged for ten years (Figures 3.7 and 3.8). But upon the tenth year, when the narrative of the Iliad starts, Agamemnon insulted god Apollo by capturing and enslaving Cressida, daughter of Cressis, priest of Apollo. Then, the god sent a nine-days plague to the Greek camp, and Agamemnon was compelled to return Cressida to her father. However, to replace Cressida, he forcibly took from Achilles his slave, Bryseis, which made Achilles so furious that he abandoned the war and withdrew, along with his troops, the famous Myrmidons. The outcome was disastrous for the Greeks (Figures 3.9–3.12).

Chapter 4
The Siege and Fall of Troy

Myriads of tales have been told of the events during the Trojan War, and the lives of many valiant warriors from both sides were lost. Paris and Menelaus duelled about Helen, and Aphrodite saved Paris just when Menelaus was about to kill him. Achilles, the greatest of the Greek warriors, slaughtered Cygnus, Troilus and many others. Odysseus and Diomedes killed thirteen Thracians, allies of the Trojans, and stole the horses of king Resus during a night raid. Ajax Telamonius fought valiantly to no avail. But the worst was that Achilles's abstention gave the Trojans great victories against the Greeks, breaking their defense lines and setting their ships on fire (Figure 4.1).

Figure 4.1 Giovanni Battista Ghisi: The Trojans repulsing the Greeks, ca. 1538.

S.A. Paipetis, *The Unknown Technology in Homer*, History of Mechanism and Machine Science 9, DOI 10.1007/978-90-481-2514-2_4, © Springer Science + Business Media B.V. 2010

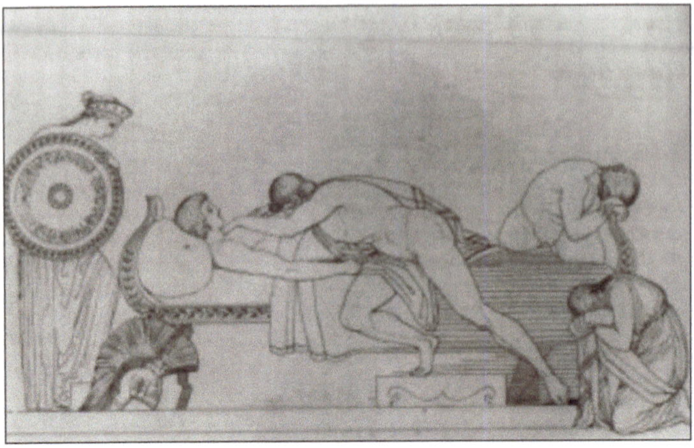

Figure 4.2 John Flaxman: Achilles weeping over Patroclus' body (illustration of Pope's translation of the Iliad, 1805).

Figure 4.3 Giorgio de Chirico, 1917: Hector and Andromache, National Gallery of Modern Art, Rome (reproduced by permission of Ministero per i Beni e le Attività Culturali).

Figure 4.4 Angelica Kauffman. Hector urging Paris to take part to the war.

Figure 4.5 The Achilles Painter, Crater, clay, red-figure painting, 510 BCE: Achilles tying the body of Hector to his chariot, The State Hermitage Museum, St. Petersburg (Photograph © The State Hermitage Museum, reproduced by permission).

Figure 4.6 Johann Balthasar Probst (1673–1748), Achilles triumphant over Hector, etching, Fine Art Museum, San Francisco (reproduced by permission).

Figure 4.7 Jacques-Louis David, 1783: Andromache weeping Hector, Pushkin Fine Art Museum, Moscow.

Achilles remained adamant with his abstention, but eventually sent his very close friend to fight in his stead, wearing Achilles's own armor. At first, Patroclus had many a success, but eventually he was killed by Hector, supported by the god Apollo. Hector managed to keep Achilles's armor as well.

Grieved by the death of his friend, Achilles (Figure 4.2) decided to return to the battle, but he was unarmed. Thetis, his mother, appealed to Hephaestus, the god-technician, to fashion a new armor for him.

Achilles returned to the war, killed Hector in a duel, tied the dead body to his chariot with a rope and dragged it around the dusty battlefield (Figures 4.3–4.7). Book 18 of the Iliad, relating these events, provides more information of military technology of the era than any other book of the Homeric Epics.

Figure 4.8 Attic clay red-painted scyphus[1] attributed to potter Brygus, 490 BCE: Priam begging of Achilles to deliver Hector's dead body to him.

Figure 4.9 Alexander Ivanov, 1824, Priam begging of Achilles to deliver Hector's dead body to him, Tretyakov Gallery, Moscow.

[1] Large drinking cup, used by Greeks and Romans, especially by poor folk.

Figure 4.10 J.-M. Moreau (1741–1814): Death of Achilles.

Achilles, on a huge funeral pyre, sacrificed a crowd of Trojan captives and organized funeral chariot races in honor of his deceased friend. King Priam of Troy visited Achilles begging of him to deliver his son's body to him (Figures 4.8 and 4.9).

Achilles eventually was killed by Paris, who fatally wounded him with an arrow in his ankle. This was the only vulnerable part of his body, because it had remained out of the waters of Stynx, while his mother was bathing him (Figure 4.10). Odysseus and Ajax Telamonius duelled over the god-made armor of the dead hero. Ajax lost and committed suicide. Paris died later, hit by an arrow launched by Philoctetes by the bow of Heracles.

Troy eventually fell thanks to a trick devised by Odysseus. A hollow Δούρειος Ἵππος (= *wooden horse*) was constructed, carrying many armed warriors inside, and left as an offering before the walls of Troy, while the Greeks pretended to have given up war and departed. To bring the huge offering into the city, the Trojans tore down part of the city walls, chose to ignore Laocoön's warning and his famous phrase: Φοβοῦ τοὺς Δαναούς καὶ δῶρα φέροντας (*beware of the Greeks even bearing gifts*[2]), as a huge sea-monster suddenly appeared and strangled Laocoön and his two sons (Figure 4.11).

Figure 4.11 Laocoön of William Blake (1826–1827).[3]

[2] In Latin: *Timeo Danaos et dona ferentes.*

[3] William Blake (1757–1827) British poet, painter, engraver and visionary mystic, who illustrated by hand a whole series of lyric and epic poems, most impressive, original and independent contributions to the western cultural tradition.

Figure 4.12 Giovanni Battista Ghisi: The fall of Troy, ca. 1540.

When the wooden horse was brought into the city, the hidden warriors emerged from their hideout and opened the gates of the city allowing the Greek army to enter and conquer Troy. Troy was totally destroyed (Figure 4.12). King Priam was put to death by Neoptolemus, son of Achilles, and infant Astyanax, Hector's son, was thrown over a wall to his death. Hecabe, Priam's spouse, Cassandra, his daughter, and Andromache, Hector's spouse, were enslaved. Helen was returned to Menelaus.

Chapter 5
Odysseus' Long Way Home

The gods considered the conquest of Troy and the destruction of its temples a sacrilege, and punished many of the Greek leaders. On the way home, the Greek fleet was almost entirely destroyed by tempests. Menelaus' ships wandered for seven years before arriving at Sparta. Agamemnon, upon returning home to Argos, was murdered by Clytemnestra, his wife, and Aegisthus, her lover, starting the ensuing horrible tragedy of the House of Atreids.

The greatest trials were reserved for Odysseus, who, pursued by the enmity of Poseidon, wandered for ten years before reaching Ithaca, having lost

Figure 5.1 J.M. William Turner (1775–1851): Odysseus deriding Polyphemus (1829), National Gallery, London (reproduced by permission).

S.A. Paipetis, *The Unknown Technology in Homer*, History of Mechanism and Machine Science 9, DOI 10.1007/978-90-481-2514-2_5, © Springer Science + Business Media B.V. 2010

Figure 5.2 Jean-Charles Cazin (1841–1901): Odysseus shipwrecked, Tate Gallery, London (reproduced by permission).

all of his companions. After leaving Troy he arrived at the land of the Cicons, then of the Lotus-eaters and then to the Cyclops, from where he fled with his companions, having blinded and tricked Polyphemus the Cyclop, son of Poseidon (Figure 5.1).

In the course of his journey, he passed by the cave of Aeolus, god of the winds, the land of Laestrygones, the islands of Circe, Calypso and of the Sirens, descended to Hades, conversed with Achilles' shadow, and then crossed the straits of Scylla and Charybdis (Figures 5.2–5.5).

Finally, passing through the meadows of Helios' cattle, he arrived at the island of the Phaeacians, where king Alcinous and his daughter Nausicaa took good care of him and sent him safe to Ithaca (Figure 5.6).

Odysseus appeared disguised in Ithaca, where, assisted by Telemachus, his son, and his faithful servants, he encountered a crowd of suitors, who, at the expense of the royal resources, were competing to claim his spouse Penelope, since Odysseus was thought to be dead (Figure 5.7). Odysseus, with unrivalled bravery and fighting ability equal to his wisdom and resourcefulness, exterminated the suitors.

Figure 5.3 Alecos Fassianos: Odysseus and the Sirens (private collection, reproduced by permission).

Figure 5.4 Alecos Fassianos: Odysseus and Calypso (private collection, reproduced by permission).

Figure 5.5 Anatole Calmems (1822–1906): Calypso.

Figure 5.6 Michele Desubleo (1602–1676): Odysseus and Nausicaa.

Figure 5.7 Pintoricchio (1454–1513): Penelope and the suitors (1509), National Gallery, London (reproduced by permission).

5.1 The Descent to Hades and the Nekyomanteion of Acheron River

In Book 10 of the Odyssey, Odysseus, in full despair, asked goddess Circe, how he would be able to know the future and whether he would ever manage to return to Ithaca. Circe recommended that he descend to the Nether World to consult the spirits of the dead and, in particular that of Teiresias, the Thebaean soothsayer:

’ὦ Κίρκη, τίς γὰρ ταύτην ὁδὸν ἡγεμονεύσει;
εἰς Ἄϊδος δ’ οὔ πώ τις ἀφίκετο νηὶ μελαίνῃ.
ὣς ἐφάμην, ἡ δ’ αὐτίκ’ ἀμείβετο δῖα θεάων·
’διογενὲς Λαερτιάδη, πολυμήχαν’ ’Οδυσσεῦ,
μή τί τοι ἡγεμόνος γε ποθὴ παρὰ νηὶ μελέσθω,
ἱστὸν δὲ στήσας, ἀνά θ’ ἱστία λευκὰ πετάσσας
ἧσθαι· τὴν δέ κέ τοι πνοιὴ Βορέαο φέρῃσιν.
ἀλλ’ ὁπότ’ ἂν δὴ νηὶ δὶ ’Ωκεανοῖο περήσῃς,
ἔνθ’ ἀκτή τε λάχεια καὶ ἄλσεα Περσεφονείης,
μακραί τ’ αἴγειροι καὶ ἰτέαι ὠλεσίκαρποι,
νῆα μὲν αὐτοῦ κέλσαι ἐπ’ ’Ωκεανῷ βαθυδίνῃ,
αὐτὸς δ’ εἰς ’Αίδεω ἰέναι δόμον εὐρώεντα.
ἔνθα μὲν εἰς ’Αχέροντα Πυριφλεγέθων τε ῥέουσιν
Κώκυτός θ’, ὅς δὴ Στυγὸς ὕδατός ἐστιν ἀπορρώξ,
πέτρη τε ξύνεσίς τε δύω ποταμῶν ἐριδούπων·
ἔνθα δ’ ἔπειθ’, ἥρως, χριμφθεὶς πέλας, ὥς σε κελεύω,
βόθρον ὀρύξαι, ὅσον τε πυγούσιον ἔνθα καὶ ἔνθα,
ἀμφ’ αὐτῷ δὲ χοὴν χεῖσθαι πᾶσιν νεκύεσσιν,
πρῶτα μελικρήτω, μετέπειτα δὲ ἡδέι οἴνῳ,
τὸ τρίτον αὖθ’ ὕδατι· ἐπὶ δ’ ἄλφιτα λευκὰ παλύνειν.
πολλὰ δὲ γουνοῦσθαι νεκύων ἀμενηνὰ κάρηνα,
ἐλθών εἰς ’Ιθάκην στεῖραν βοῦν, ἥ τις ἀρίστη,
ῥέξειν ἐν μεγάροισι πυρήν τ’ ἐμπλησέμεν ἐσθλῶν,
Τειρεσίῃ δ’ ἀπάνευθεν ὄϊν ἱερευσέμεν οἴῳ
παμμέλαν’, ὃς μήλοισι μεταπρέπει ὑμετέροισιν.
αὐτὰρ ἐπὴν εὐξῇσι λίσῃ κλυτὰ ἔθνεα νεκρῶν,
ἔνθ’ ὄϊν ἀρνειὸν ῥέζειν θῆλύν τε μέλαιναν
εἰς ῎Ερεβος στρέψας, αὐτὸς δ’ ἀπονόσφι τραπέσθαι
ἱέμενος ποταμοῖο ῥοάων· ἔνθα δὲ πολλαὶ
ψυχαὶ ἐλεύσονται νεκύων κατατεθνηώτων.

And who shall guide me upon this voyage – for the house of Hades is a port that no ship can reach.
You will want no guide, she answered; raise you mast, set your white sails, sit quite still, and the North Wind will blow you there of itself. When your

> ship has traversed the waters of Oceanus, you will reach the fertile shore of Proserpine's country with its groves of tall poplars and willows that shed their fruit untimely; here beach your ship upon the shore of Oceanus, and go straight on to the dark abode of Hades. You will find it near the place where the rivers Pyriphlegethon and Cocytus (which is a branch of the river Styx) flow into Acheron, and you will see a rock near it, just where the two roaring rivers run into one another.
>
> When you have reached this spot, as I now tell you, dig a trench a cubit or so in length, breadth, and depth, and pour into it as a drink-offering to all the dead, first, honey mixed with milk, then wine, and in the third place water-sprinkling white barley meal over the whole. Moreover you must offer many prayers to the poor feeble ghosts, and promise them that when you get back to Ithaca you will sacrifice a barren heifer to them, the best you have, and will load the pyre with good things. More particularly you must promise that Teiresias shall have a black sheep all to himself, the finest in all your flocks.
>
> When you shall have thus besought the ghosts with your prayers, offer them a ram and a black ewe, bending their heads towards Erebus; but yourself turn away from them as though you would make towards the river. On this, many dead men's ghosts will come to you ...
>
> (Od. 10.501–530)

At the exact place that Circe suggested, near the banks of Acheron River (= *River of Woe*) in Thesprotia of Epirus, Western Greek mainland, lies what in ancient times was believed to be the entrance to Hades or the Underworld, probably due to Acheron flowing through dark gorges, going underground at several places. According to mythology, upon death, a soul was lead by Hermes to the entrance of the underworld to cross the Acheron. There, a single ferry, ran by Charon, carried the soul across the river. A fare was required, therefore, a coin was placed on the lips of the dead upon burial. Those who could not pay remained forever trapped between the two worlds.

On the bank of Acheron is the Nekyomanteion or Necromanteion (νέκυς = νεκρός = *dead man*), an oracle, in which necromancy was practiced. Necromancy, or communication with the dead, usually served divination purposes, an activity current in ancient times among Assyrians, Babylonians, Egyptians, Greeks, Romans, and Etruscans, as well as in medieval Europe.

The ruins of Acheron Nekyomanteion and the surrounding area are well preserved. Excavations have revealed much information about the conditions under which individuals were prepared to communicate with their departed relatives.[1] A candidate, after several days of fasting and sleeplessness, being

[1] S.I. Dakaris, (a) Das Taubenorakel von Dodona und das Totenorakel bei Ephyra, *AntK* (1963), beih. 1, pp. 35–54. (b) Ephyra, in *Princeton Encyclopedia of Classical Sites*, Prince-

Figure 5.8 Chamber of the dead, Nekyomanteion of Acheron River, as it is today (from author's collection).

fed with intoxicating food, was lowered,[2] to the underground Chamber of the Dead for the scheduled meeting (Figure 5.8). The priesthood employed sophisticated means, in order to convince the candidate of the authenticity of his experience.

In staging this ritual, sound effects obviously played an important role, and the construction of the Chamber indicates that its designers possessed important knowledge of acoustics: A series of stone arches, placed at regular intervals along the tunnel, provide evidence that sounds were transmitted in a specific way serving the purpose of the alleged communication. By using the method of Boundary Elements (BE), a thorough numerical investigation was performed (Figure 5.9), and the results revealed a lot on the operation of the structure.[3]

The study led to the following results:

ton University Press, 1976, pp. 310–311, (c) *The Nekyomanteion of the Acheron*, Archaeological Receipts Fund, Ministry of Culture, Athens, 1993.

[2] The remnants of a hoisting mechanism or catapult of the 3rd century are exhibited at the Archaeological Museum of Ioannina.

[3] S.A. Paipetis, D. Polyzos, J.P. Agnantiaris and E.J. Sellountos, Acoustic Behaviour of the Chamber of Dead in the Necromanteion of Acheron River, in *Extraordinary Machines and Structures in Antiquity*, S.A. Paipetis (Ed.), Peri Technon Publ., Patras, Greece, 2003.

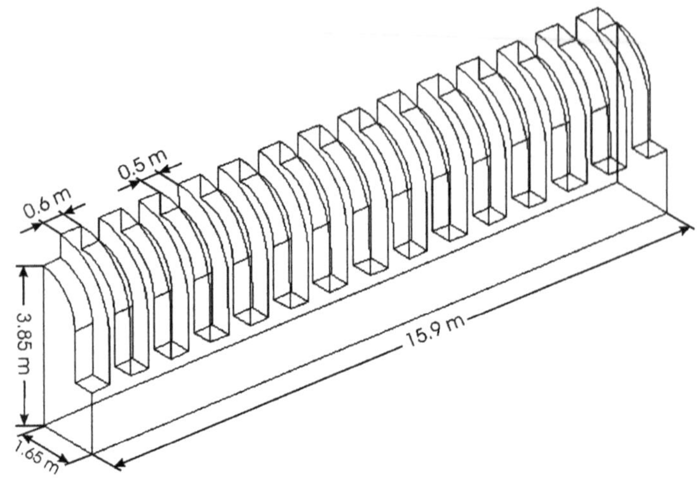

Figure 5.9 BE model for the acoustic study of the Chamber.

(a) High sound pressures develop at surfaces between two adjacent ribs. This means that, with (low) frequencies of 100 and 200 Hz, the Chamber behaves as an anechoic chamber, i.e., a considerable amount of the sound energy is reflected on the walls and entrapped between the ribs and eventually attenuated. Therefore, at these frequencies the sound quality should be remarkably high. The results could be more impressive if one considers that the floor is not rigid but a sound-absorbing surface.
(b) The anechoic behavior of the Chamber would be lost, if the walls had smooth surfaces, e.g., no ribs.
(c) The sound pressure near the end-cups and at the center of the Chamber undergoes much higher amplification and the sound is of higher quality than with the no-ribs chamber. Also, on both end-cups the amplitude of the sound pressure is the same in this range of frequencies.

The present investigation was by no means exhaustive, and it is expected that further research may provide more interesting results on the specific functions of the monument.

Chapter 6
Trojan War and Cultural Tradition

In the western cultural tradition no story, exempting the Bible, has inspired more artists, writers, painters, sculptors and so on, than the Trojan War. This is evident from the masterpieces of paintings and sculptures already presented in the previous chapters, representing a small part only of the tradition developed through the centuries. It is remarkable that various artists place the events of the Trojan War in their own environment – country, landscape, architecture, human figures and clothing, showing that partial narratives composing the whole, have been incorporated in the traditions of various peoples. It also explains why these stories have not remained static but are constantly evolving with time, and no version can be defined as authentic, as for example is the case with the Bible.

The extant works, complete or in fragments, of the great tragedians of ancient Greece, were based to a great extent on the events of the Trojan War and, in particular, on those that followed the fall of Troy: Numerous works of Aeschylus (the *Oresteia* trilogy, *Agamemnon*, *Choeforoi*, *Eumenides*), of Sophocles (*Ajax*, *Philoctetes*) and of Euripides (*Troads*, *Hecabe*, *Helene*, *Andromache*, *Cyclops*, *Iphigeneia in Tauris*, *Iphigeneia in Aulis*) are famous examples. Also, the Greek philosophers have dealt, one way or another, with the same subject, while historians, like Herodotus and Thukydides, have analyzed the causes and the events of the war in depth. Alexander the Great always carried along a copy of the Iliad.

The story of the war was adopted by the Romans too. The famous *Aeneid* of Virgil tells the story of Aenaeas, a Trojan prince, who, after the fall of Troy, fled away with his family and found a refuge in Italy, where he established *Lavinium*, the main centre of the Latin connection from which the Roman people originate. In the Middle Ages, from the Renaissance till our days, writers have kept telling the same old story again and again. One meets

S.A. Paipetis, *The Unknown Technology in Homer*, History of Mechanism and Machine Science 9, 49
DOI 10.1007/978-90-481-2514-2_6, © Springer Science + Business Media B.V. 2010

(a)

(b)

Figure 6.1 Trojan War and humour: Two incomparable compositions of Arcas, one of the wittiest modern cartoonists in Greece, dealing with ... life in Paradise (reproduced by permission). (a) Background screams: "Idiot!", "You are an idiot!" The man-in-Paradise asks: "Angel, what is all this screaming?" The angel replies "The Trojans are arguing again. I am fed up with them! ... Three thousand years the same story!" But the screams continue: "Who said, 'OK, they are gone'?" and the reply: "Yes, but who said, 'O, what a beautiful horse'?", (b) The angel says: "Here is Achilles with the fatal arrow". The man-in-Paradise asks: "But was not he hit at the ankle?", "No, at the breastbone, but Homer's handwriting was awful!" There is a double catch here: The words "πτέρνα" (= *ankle*) and "στέρνον" (= *breastbone, sternum*) sound and also look, in writing, almost the same in Greek. Besides the pun: Homer's "bad handwriting" has created this historical confusion, as it refers to a time that writing was not supposed to be yet invented!

Odysseus and Diomedes in Dante's *Hell* again. Shakespeare writes *Troilus and Cressida*, while modern playwrights perpetuate the tradition: J.P. Sartre (*The Flies*), E. O'Neal (*Mourning Becomes Electra*), J. Giraudoux (*A Tiger at the Gates*), James Joyce (*Ulysses*), Nikos Kazantzakis (*Odyssey*, after the return to Ithaca, consisting of 33,333 verses), C. Cavafy, T.S. Eliot, W.H. Auden, Arthur Clark in *2001: A Space Odyssey, The Adventures of Sinbad* from *Thousand and One Nights of Arabia*, the film *O Brother, Where Art Though?* by Joel Ethan Coen, the science fiction work *Space Chantey* by R.A. Lafferty, the volume *Odyssey* of the band *Progressive Metal*, the

Figure 6.2 Achilleion Palace: Main entrance (top) and the Garden of the Muses (bottom).

Japanese cartoon film *Odyssey 31* and numerous others. That is, old and modern cinema, music, ballet and humour have been influenced by Homer (Figure 6.1).

Moreover, the numbers of toponyms, trade-names of commercial businesses and logos of companies and unions, names of special software and

(a)

(b)

Figure 6.3 Two beautiful bronze plaques on either side of the palace entrance (House of Caponetti, Naples): (a) Helios with his four-horse chariot and (b) Zeus launching his thunderbolts against the Titans.

countless more, inspired by the Trojan War, is astonishing. Finally, recent discoveries relate the story of the Trojan War, as it appears in the Homeric Epics, with modern science and technology, in fact, in a most surprising way, as will be expounded in the following section.

6.1 An Architectural Masterpiece in Honour of Achilles

Achilleion Palace is a magnificent edifice in Corfu Island, in north-western Greece. It was built upon order of Empress Elisabeth of Austro-Hungary between 1889 and 1891 by the Italian architect Rafaelo Carito, in the architectural style of Pompeii, including also elements of Ionic, Roman and Aeolic traditions (Figure 6.2).[1] Achilleion Palace is now public property and is used as a Museum.

The whole Palace reflects the admiration of Empress Elisabeth (or Sissy, well-known from romantic films) for Greek Mythology in general and for Achilles in particular. Numerous depictions of gods, goddesses, heroes, Titans and other mythological figures in the form of sculptures, paintings, frescos etc. exist all over the building and the gardens, as for example in Figure 6.3.

Many art objects created by famous artists of the time are related to scenes of the Iliad and the Odyssey: Figure 6.4 presents the first encounter of shipwrecked Odysseus with princess Nausicaa of the Phaeacians. It is said that the artist was inspired by the following verses:

---

[1] All images are reproduced by permission.

Figure 6.4 Ludwig Thiers, Odysseus and Nausicaa (1859), Achilleion Palace, Elisabeth's private apartments.

Figure 6.5 Franz Josef Karl, Edler von Matsch, Achilles Triumphant, at the exquisite main stair-case.

Figure 6.6 Ernst Gustav Herter, Achilles Dying, Berlin 1884.

σοί δὲ θεοὶ τόσα δοῖεν ὅσα φρεσὶ σῇσι μενοινᾷς,
ἄνδρα τε καὶ οἶκον, καὶ ὁμοφροσύνην ὀπάσειαν
ἐσθλήν· οὐ μὲν γὰρ τοῦ γε κρεῖσσον καὶ ἄρειον,
ἢ ὅθ' ὁμοφρονέοντε νοήμασιν οἶκον ἔχητον
ἀνὴρ ἠδὲ γυνή.

As for you, may gods grant all your heart desires – may they give you a husband, home, and mutual harmony, a noble gift – for there is nothing better or a stronger bond than when man and wife live in a home sharing each other's thoughts.

(Od. 6.180–184)

Figure 6.5 shows a magnificent oil painting by Franz Matsch[2] (1892), covering the whole wall of the upper part of the superb main staircase. This great work is the centre of interest in the palace, also because of the stories accompanying it: The painter is supposed to have committed suicide, because he failed to show the motion of the wheel of Achilles's chariot while dragging Hector's body, which appears fully static. In fact, this may correspond to the particular style the painter has chosen.

[2] Franz Josef Karl, Edler von Matsch (1861–1942), Austrian painter and a representative of *Art Nouveau*, an art movement of the end of 19th century and the beginning of the 20th.

Figure 6.7 Johannes G. Götz,[4] Achilles Victor.

Finally, Figure 6.6 shows *Achilles Dying*, one of the most famous works of Ernst Gustav Herter.[3]

Elisabeth was assassinated in 1898 in Geneva, and the Palace remained empty until 1907, when it was bought by Kaiser Wilhelm II of Germany, who made a lot of changes. One of his additions was the huge bronze statue of *Achilles Victor* (Figure 6.7), the biggest statue of Achilleion with unreal dimensions (5.5 m high or 11.5 m along with its base and the spear) and 4.5 ton weight. The statue was Kaiser's counterbalance against the masterpiece of *Achilles Dying*. The inscription Τόνδε 'Αχιλλῆα Πηλείδην Γερμανῶν Κραταιῶν Γουλιέλμος στῆσαι, μῆμα ἐπιγιγνωμένοις (*To Achilles, son of Peleus, Wilhelm of Puissant Germans has erected in remembrance for the generations to come*) was removed by the French, who, at a later stage, seized the island.

[3] Ernst Gustav Herter (1846–1917), famous German sculptor, who lived and worked in Berlin. He specialized in statues of mythological figures.

[4] Johannes Gottfried Götz (1865–1934) German sculptor.

Chapter 7
Scientific Knowledge in the Homeric Epics

> Under the pillow of the modest bed of Asia's conqueror,[1] put the wise Stageirite[2] the golden Iliad, so that, even by phantasy, his noble disciple dream nicely, but also who of the ancient leaders, poets, artists, orators, historians, geographers did not have the immortal epics of Homer as a vademecum, guide and exemplary in his science, in his art, in life's tribulations?
>
> K.D. Zeggelis[3]

According to an attitude prevailing in the past, the ancient Greeks were unable to work out technological achievements. For example, Carl Jung,[4] maintaining that technological skills were due to the development of "directed thinking", to his view a product of the social conditions of mediaeval times, states:[5]

> Directed thinking or, as we might also call it, thinking in words, is manifestly an instrument of culture, and we shall not be wrong in saying that the tremendous work of education which past centuries have devoted to directed thinking, thereby forcing it to develop from the subjective, individual sphere to the objective, social sphere, has produced a readjustment of the human mind to which we owe our modern empiricism and technics. These are absolutely new developments in the history of the world and were unknown to earlier ages. *Inquiring minds have often wrestled with the question of why the first-rate*

[1] Alexander the Great.

[2] Aristotle, born in the city of Stageira, Macedonia, Northern Greece.

[3] K.D. Zeggelis, *The Science of Nature in Homer*, Athens, 1891, republished in 1987 by University of Patras Editions, introductory essay by S.A. Paipetis (in Greek).

[4] Carl Gustav Jung (1875–1961): Swiss psychiatrist and founder of analytical psychology.

[5] C.G. Jung: *Symbols of Transformation*, Ch. 2, *Two Kinds of Thinking*, Routledge and Keagan Paul, London, 1967.

S.A. Paipetis, *The Unknown Technology in Homer*, History of Mechanism and Machine Science 9, DOI 10.1007/978-90-481-2514-2_7, © Springer Science + Business Media B.V. 2010

> *knowledge which the ancients undoubtedly had of mathematics, mechanics, and physics, coupled with their matchless craftsmanship, was never applied to developing the rudimentary techniques already known to them (e.g., the principles of simple machines) into a real technology in the modern sense of the word, and why they never got beyond the stage of inventing amusing curiosities.* There is only one answer to this: the ancients, with a few illustrious exceptions, entirely lacked the capacity to concentrate their interest on the transformations of inanimate matter and to reproduce the natural process artificially, by which means alone they could have gained control of the forces of nature. What they lacked was training in directed thinking. The secret of cultural development is the mobility and disposability of psychic energy. Directed thinking, as we know it today, is a more or less modern acquisition which earlier ages lacked.

Jung's way of thinking is getting clearer through his following reference to myth, mainly to the ancient Greek myth, included in the same text:

> All the creative power that modern man pours into science and technics the man of antiquity devoted to his myths. This creative urge explains the bewildering confusion, the kaleidoscopic changes and syncretistic regroupings, the continual rejuvenation, of myths in Greek culture. We move in a world of fantasies which, untroubled by the outward course of things, well up from an inner source to produce an ever-changing succession of plastic or phantasmal forms. This activity of the early classical mind was in the highest degree artistic: the goal of its interest does not seem to have been how to understand the real world as objectively and accurately as possible, but how to adapt it aesthetically to subjective fantasies and expectations.

Jung's distinction between the "modern man" and the "man of antiquity" is a source of heavy error, at least as far as ancient Greece is concerned. As a point of temporal and historical reference, he choose central Europe of 19th and 20th centuries, who had to go through the incredible terror of the Middle Ages as well as of wars and revolutions, to arrive eventually at the modern democratic kind of government. But, as it is well known, all this pre-existed in the classical antiquity. In the same way that in the mind of the Greek of antiquity Aristotle's logic and Plato's metaphysics co-existed. And, as far as the specific subject is concerned, it is obvious that "in parallel to the important artistic activity of the mind, the latter was also aiming at conceiving the essence of real world as objectively and accurately as possible". At least, this is proved by the most clearly scientific theories of pre-Socratic, etc., philosophers, which essentially contain the seeds of modern scientific thought, as well as by the great technological achievements of Archimedes, Thales, Eupalinus and other engineers of ancient times. The inherent interest for technology of the ancient Greeks is demonstrated by the presence of a god-technician, Hephaestus, who, as shown in the sequence, was an engineer-designer in the modern sense of the word.

Finally,in a footnote, Jung adds the following:

> There was as a matter of fact no external compulsion which would have made technical thinking necessary. The labour question was solved by an endless supply of cheap slaves, so that efforts to save labour were superfluous. We must also remember that the interest of the man of antiquity was turned in quite another direction: he reverenced the divine cosmos, a quality which is entirely lacking in our technological age.

This thought is particularly simplistic, since, in essence, maintains that, in the antiquity, due to the existence of slaves, energy was available in great abundance and, consequently, there were no motives to develop new techniques, either aiming at the production of mechanical energy from heat or at the more efficient and useful utilization of the existing energy sources. But this is in conflict with a fundamental principle of economy, that human needs are unlimited. On the other hand, there is no evidence of prime movers available, e.g., engines capable of converting thermal energy into mechanical work in antiquity.

Even if the said view were true in times of peace, it was certainly untrue in war times, where the winner-to-be should possess a bigger and better equipped and trained army than his opponents. No war can be conducted by armies of slaves.

Relevant to this is the apotheosis of a healthy and well trained body, expressed by the glorification of athletic spirit, associated with the principle of "fair contest" of the Olympic Games.[6]

If such attitudes were prevailing in classical antiquity, the more was it natural to prevail at much earlier times, to which the Homeric Epics refer. Eventually the view was accepted that a detailed analysis of the Homeric Epics from the point of view of elements of knowledge of each individual science would be a difficult and purposeless task. This view seems now in need of re-thinking. Of course, notwithstanding the inherrent inability of an epic to provide correct and detailed accounts of things and events, the unavoidable mixing of old and new, the difficulties in articulating various concepts accurately from the etymology of terms, etc., which are real problems hard to solve indeed.

However, Homer provides accounts on many interesting machines, including even automation and artificial intelligence. Certainly, he does not

---

[6] It is well known that the home-towns of Olympic winners used to tear down their defensive walls, to emphasize that with such worthy men, they were useless.

describe, as stated, prime movers, although, in many cases, their existence is a prerequisite for other, very advanced devices. Many believe that Homer meant that such engines did indeed exist, while others consider this view totally unacceptable. However, some of the attributes described may be poetic conceptions and it must be emphasized that those Homeric accounts are substantially important, since:[7]

(a) They show that, in the Homeric era, the idea of an automat performing practical work, even create structures, was not considered impossible at all.
(b) They refer to real, practical needs for automation. Such needs are usually the catalyst for technological development.

But also, besides structures and devices, on many accasions, laws of nature are formulated in view of specific applications. This means that a solid background of knowledge existed in Homeric times, approaching the limits of scientific knowledge, and was not just the results of observations or long-term practical experience. Such examples are treated in the chapters to follow.

In fact, in specific cases, mechanical structures may be subject to structural analysis by means of modern analytical and/or numerical tools, as well as of computer codes. These analyses have shown that, for the design of such structures, like the shields of Achilles and Ajax, fully modern methods and products are necessary.[8]

These various cases are not obvious to non-specialists, which leads to the conclusion that research on the Homeric Epics must be an interdisciplinary operation, i.e., a synergy of specialists from several fields of science and technology, conducted in depth of time, following scientific progress and the discovery of new ways of research and development.

At this point, it is worth mentioning that the first systematic attempt to detect and record elements of scientific knowledge in the various fields, contained in the Homeric Epics, was made by the late Constantine D. Zeggelis, Professor of the Technical University of Athens and Senator, in a book, entitled "The Science of Nature in Homer" (1891). The book in camera-copy form, along with an extended introduction by the present author was re-published by University of Patras Editions (1987).

[7] A.D. Dimarogonas, "Machines in Homer", in *Extraordinary Machines and Structures in Antiquity*, S.A. Paipetis (Ed.), Peri Technon Publ., Patras, 2004.

[8] S.A. Paipetis and V. Kostopoulos, "Defensive Weapons in Homer", in *Extraordinary Machines and Structures in Antiquity*, S.A. Paipetis (Ed.), Peri Technon Publ., Patras, 2004.

Chapter 8
On Science and Technology

Science (in Greek ἐπιστήμη (= epistēmē) from the verb ἐπίσταμαι (= *to know well*, in Latin *scientia*, from the verb *scire* = to know) is any system of knowledge dealing with the physical world and its phenomena and presupposes unbiased observation and experimentation. In general, all sciences search for knowledge covering generalized truths or the operation of fundamentals laws of nature.

Figure 8.1 The dawn of technology: The ape-man is using a long bone both as a tool and a weapon. From the classical masterpiece of Stanley Kubrick *2001: A Space Odyssey*, based on the homonymous fiction by Arthur Clark.

S.A. Paipetis, *The Unknown Technology in Homer*, History of Mechanism and Machine Science 9, DOI 10.1007/978-90-481-2514-2_8,

The birth of science is trivially related to the ceaseless efforts of Man to solve problems connected to his survival. However, besides this, it is a fact that the search for truth is an innate human quality fully interwoven with the very existence and substance of Man. This is expressed admirably through the etymology of the word ἄνθρωπος (*anthropos*, i.e., the Man) in Plato's *Cratylus*, as originating from the phrase ἀναθρόω ἅ ὄπωπα (= *to reflect on what I have observed*), which, in fact, expresses the foundations of scientific methodology. So, even with all of his survival problems solved, Man would never cease to aspire to new knowledge (see also Aristotle: "φύσει τοῦ εἰδέναι ὀρέγετ' ἄνθρωπος", i.e., "by nature man likes knowledge"). Undoubtedly, this quest becomes more intense, or even imperative, under pressure, as, for example, with war, when the search for "useful" or "applicable" knowledge becomes most urgent. In our days, this is expressed by the distinction between science and technology, or by classifying research into *basic* or *fundamental*, *applied*, and *technological* or *industrial* research.

Technology is the application of scientific knowledge for practical purposes in human life, or, as sometimes stated, to change and control human environment (Figure 8.1).

However, irrespective of its origin, science has a fundamental attribute: As opposed to art, where creation is the result of the efforts of one sole person,[1] science and technology are produced by collective effort or team work. Maybe this was meant by Isaac Newton's famous quote "If I have seen a little further, it is because I was standing on the shoulders of giants", in the sense of the cumulative property of science, e.g., every new achievement is a further development of what already exists. Scientific and technological developments proceed at an increasing rate, since with increasing background of knowledge and technological achievements, science and technology may evolve much faster.

Therefore, this is how technology has developed and keeps on developing ceaselessly in time. From the simple machines, which increased by a manifold the physical strength of humans or animals, it went on to prime movers through the Industrial Revolution of the 18th century. And, finally, to the modern revolution of Informatics, whose prospects are unlimited. As stated, the formal definition is that technology is *the application of scientific knowledge in methods and activities, by which Man can manage or modify his environment.*[2]

---

[1] Bernard Dixon and G.S. Holister, *Ideas of Science*, Basil Blackwell, Oxford, 1984, p. 1, Art is "I", science is "we" (G.S. Holister).

[2] In a most disastrous way unfortunately.

Respectively, science is defined as *the effort of Man to conceive and comprehend the world.* It was created much later, by the development of the great civilizations, i.e., after 3,000 BCE. Science and technology moved along separate tracks, which were crossed at the point of common use of the methods of mathematics and physics. The cross-fertilization of the two has led to new processes, by means of which inventiveness and creativity of Man have produced admirable achievements. Scientific research, in its three distinct forms (basic, applied, technological) extends from the elementary effort to increase human knowledge up to specific applications and its utilization for the benefit of Man, something not always "beneficial" and not always safe. Accordingly, a purely scientific problem of yesterday develops into a technological problem of today and eventually to an industrial or military application of tomorrow. In that sense, the practically unified dipole "Science and Technology" is characterized by certain specific properties:

1. They are cumulative processes: Every new piece of knowledge is based on pre-existing knowledge, which, in general, is the result of team work (as opposed to art objects, which, by nature, are creations of individuals.)
2. With increasing knowledge background, new knowledge is acquired faster and easier.
3. If, for some reason, the existing knowledge disappears, the whole process of acquisition of new knowledge must restart from zero.
4. Every new scientific achievement creates new possibilities for all kinds of technical, financial and social developments, new conditions are generated, not existing before, or it makes existing conditions to evolve faster, easier and at lower cost.
5. A very advanced technology may constitute a threat for humanity, not only because of mass destruction weaponry, but also due to the constantly increasing consumption of energy and raw materials, necessary to sustain and develop further existing ways of life. A huge number of life forms keep on disappearing, while respective numbers are classified as endangered species, humans included on some occasions.

On the other hand, it is evident that a disruption of this collective process, not just by a disastrous war, by which, as stated, evolutionary processes are intensified, but for much more serious reasons, as it is, for example, the complete destruction of a civilization due to natural or other causes, by which everything must start again from the beginning. May be this has happened more than once in the history of our planet.

This last hypothesis makes all kinds of "lost civilizations" a very attractive theme for research scientists, as well as for science fiction writers. In fact, many nonsensical theories were developed in scientific disguise. However, many of the surviving monuments of the past, such as the Pyramids of Egypt, keep providing solid bases for hypotheses on extraordinary knowledge possessed by their creators. A thorough investigation extended deeply in the past, e.g., in past geological periods, is not always possible: lots of evidense may be burried under the glaciers or the rain forests, covering a great part of earth surface, which, several thousands of years ago, might be habitable areas with mild climate.

On the other hand, extant pieces of literature, mainly poetry, concerning prehistory, although often giving rise to respective thoughts, usually they are written in strongly symbolic language. Therefore, from the respective accounts, one cannot easily distinguish between real situations and fantasy or even artistic conceptions, which, however, by inspiration, may approach truth in a more direct way than costly and laborious scientific research.

On many occasions, accounts are accurate and realistic and refer to themes insinuating possession of advanced knowledge and extraordinarily developed technology. Even the remnants of words, along with their etymology, may constitute interesting testimonies. Therefore, before the eyes of individual specialists, amazing information appears, potentially suggesting the existence of technologically advanced civilizations in prehistoric times. Undoubtedly, the Homeric Epics, Iliad and Odyssey, can be the object of such a study.

Part 2
Principles of Natural Science

Chapter 9
Chariot Racing and the Laws of Curvilinear Motion

9.1 The Mycenaean Chariot

In the early Bronze Age, land transportation was effected with wooden vehicles pulled by animals. A model gharry from Crete is dated from 2,000 BCE or earlier. Such gharries had whole-body, rigid wheels and were pulled by oxen. Around that time, horses appear in Crete, originating from the East, and are depicted in early Cretan seal-rings.

A light chariot with spoked wheels was evidently developed in Syria or Northern Mesopotamia at about the beginning of the 2nd millennium BCE and quickly propagated all over Middle East as military vehicle. Such chariots appear engraved in tombstones of Mycenaean vaulted graves, as well as in Cretan seal-rings of about 1450 BCE. Besides in war, they were used in hunting, as well as for traveling. This is the only military chariot appearing in the Iliad. Roads for wheeled vehicles were constructed at around the end of the Bronze Era, especially bridging streams and rivers, remnants of which can still be found in Mycenae area.

Chariots of Mycenaean and Archaic Greece had very light and flexible wheels, made by bending of very thin wood – willow, elm or cypress – usually with four spokes only (Figures 9.1 and 9.2). Such a wheel was very elastic and acted as spring suspension, allowing the chariots to trot on the rough ground of the Greek hillside, where heavier and more rigid vehicles would be useless. In fact, the wheel's hub was bending as a bow under the chariot's weight. Four-wheel chariots of similar design were developed at later times (Figure 9.3).

S.A. Paipetis, *The Unknown Technology in Homer*, History of Mechanism and Machine Science 9,
DOI 10.1007/978-90-481-2514-2_9,

(a) (b)

Figure 9.1 Mycenaean military chariot. (a) Sculpture of the 12th century BCE). (b) The Leagros Group, Hydria, Clay; Black-figure painting (circa 550 BCE), The State Hermitage Museum, St. Petersburg (Photograph © The State Hermitage Museum, reproduced by permission).

Figure 9.2 Reconstruction of a Mycenaean chariot (1400–1200 BCE). The Homeric chariot was very flexible, being constructed by bending of very thin wood.

The art of road construction was brought to excellence by the Romans[1] (Figure 9.4). Road grooves can provide reliable information on the construction and attributes of Roman chariots (Figure 9.5).

[1] L. Corradi et al., Building Techniques of Roman Roads: The Via Flaminia from Narni to Forum Flaminium in Umbria, in *Extraordinary Machines and Structures in Antiquity*, S.A. Paipetis (Ed.), Peri Technon Publications, Patras, Greece, 2002.

Figure 9.3 Vase from about the end of Dark Ages. At the lower zone, horse-driven chariots at a funeral procession, reflecting Mycenaean practices, are depicted.

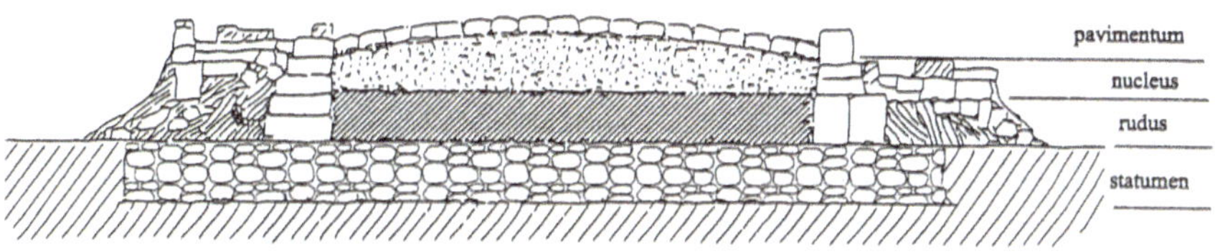

Figure 9.4 Typical cross section of the Roman via publica.

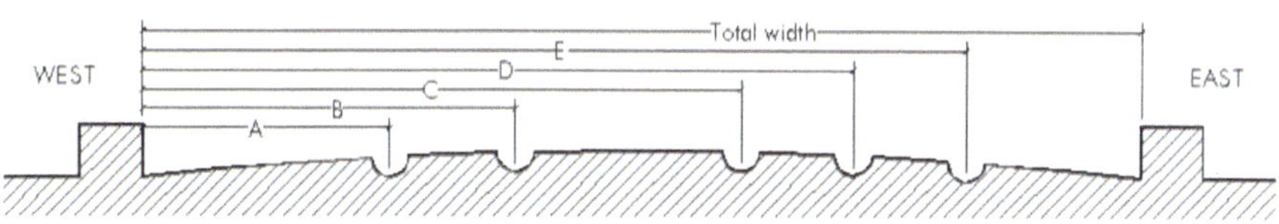

Figure 9.5 Survey of road grooves on the paving of Via Flaminia in Carsulae.

Actually, the present chapter is dealing with fast-moving chariots, used in war or in chariot races, where the capability of the vehicles to execute manoevres at high speed is important. Especially, in case of quick, abrupt turns, the laws of Dynamics take effect, and inertial forces develop, potentially threatening not just the stability of the moving structure, but also the outcome of a battle or a game and the life of the charioteer. A brilliant account of a chariot race, in which the laws of curvilinear motion of rigid bodies are accurately formulated, is given in the Iliad.

9.2 Nestor's Instructions to Antilochos

In the Iliad, Book 23, Nestor, King of Pylos, appears to instruct Antilochos, his son, how to win the chariot race, organized by Achilles in honour of his dear friend Patroclus, who died in a duel with Hector, the Trojan Prince:

> Ἀντίλοχ', ἤτοι μὲν σε νέον περ ἐόντ' ἐφίλησαν
> Ζεὺς τε Ποσειδάων τε, καὶ ἱπποσύνας ἐδίδαξαν
> παντοίας· τῶ καὶ σε διδασκέμεν οὔ τι μάλα χρεώ·
> οἶσθα γὰρ εὖ περὶ τέρμαθ' ἑλισσέμεν· ἀλλὰ τοι ἵπποι
> βάρδιστοι θείειν· τῷ τ' οἴω λοίγι' ἔσεσθαι.
> τῶν δ' ἵπποι μὲν ἔασιν ἀφάρτεροι, οὐδὲ μὲν αὐτοὶ
> πλείονα ἴσασιν σέθεν αὐτοῦ μητίσασθαι.
> ἀλλ' ἄγε δὴ σὺ φίλος μῆτιν ἐμβάλεο θυμῶ
> παντοίην, ἵνα μή σε παρεκπροσφύγῃσιν ἄεθλα.
> μήτι τοι δρυτόμος μέγ' ἀμείνων ἠὲ βίηφι·
> μήτι δ' αὖτε κυβερνήτης ἐνὶ οἴνοπι πόντῳ
> νῆα θοὴν ἰθύνει ἐρεχθομένην ἀνέμοισι·
> μήτι δ' ἡνίοχος περιγίγνεται ἡνιόχοιο.
> ἀλλ' ὅς μὲν θ' ἵπποισι καὶ ἅρμασιν οἷσι πεποιθὼς
> ἀφραδέως ἐπὶ πολλὸν ἑλίσσεται ἔνθα καὶ ἔνθα,
> ἵπποι δὲ πλανόωνται ἀνὰ δρόμον, οὐδὲ κατίσχει·
> ὅς δὲ κε κέρδεα εἰδῇ ἐλαύνων ἥσσονας ἵππους,
> αἰεὶ τέρμ' ὁρόων στρέφει ἐγγύθεν, οὐδέ ἑ λήθει
> ὅππως τὸ πρῶτον τανύσῃ βοέοισιν ἱμᾶσιν,
> ἀλλ' ἔχει ἀσφαλέως καὶ τὸν προὔχοντα δοκεύει.
> σῆμα δὲ τοὶ ἐρέω μάλ' ἀριφραδές, οὐδὲ σὲ λήσει.
> ἕστηκε ξύλον αὖ ὅσον τ' ὄργυι' ὑπὲρ αἴης
> ἢ δρυὸς ἢ πεύκης· τὸ μὲν οὐ καταπύθεται ὄμβρῳ,
> λᾶε δὲ τοῦ ἑκάτερθεν ἐρηδέδαται δυὸ λευκὼ
> ἐν ξυνοχῇσιν ὁδοῦ, λεῖος δ' ὁ ἱππόδρομος ἀμφὶς
> ἤ τευ σῆμα βροτοῖο πάλαι κατατεθνηῶτος,
> ἢ τό γε νύσσα τέτυκτο ἐπὶ προτέρων ἀνθρώπων,

καὶ νῦν τέρματ’ ἔθηκε ποδάρκης δῖος ’Αχιλλεύς.
τῷ σὺ μάλ’ ἐγχρίμψας ἐλάαν σχεδὸν ἅρμα καὶ ἵππους,
αὐτὸς δὲ κλινθῆαι ἐϋπλέκτῳ ἐνὶ δίφρῳ
ἧκ’ ἐπ’ ἀριστερὰ τοῖιν· ἀτὰρ τὸν δεξιὸν ἵππον
κένσαι ὁμοκλήσας, εἶξαὶ τὲ οἱ ἡνία χερσίν.
ἐν νύσσῃ δὲ τοι ἵππος ἀριστερὸς ἐγχριμφθήτω,
ὡς ἄν τοι πλήμνη γε δοάσσεται ἄκρον ἱκέσθαι
κύκλου ποιητοῖο· λίθου δ’ ἀλέασθαι ἐπαυρεῖν,
μὴ πως ἵππους τε τρώσῃς κατὰ θ’ ἅρματα ἄξῃς·
χάρμα δὲ τοῖς ἄλλοισιν, ἐλεγχείη δὲ σοὶ αὐτῷ
ἔσσεται· ἀλλὰ φίλος φρονέων πεφυλαγμένος εἶναι.
εἰ γὰρ κ’ ἐν νύσσῃ γε παρεξελάσῃσθα διώκων,
οὐκ ἔσθ’ ὅς κέ σ’ ἕλῃσι μετάλμενος οὐδὲ παρέλθῃ,
οὐδ’ εἴ κεν μετόπισθεν ’Αρίονα δῖον ἐλαύνοι
’Αδρήστου ταχὺν ἵππον, ὅς ἐκ θεόφιν γένος ἦεν,
ἤ τοὺς Λαομέδοντος, ὅ ἐνθάδε γ’ ἔτραφεν ἐσθλοί.

Antilochus, said Nestor, you are young, but Jove and Neptune have loved you well, and have made you an excellent horseman. I need not therefore say much by way of instruction. You are skilful at wheeling your horses round the post, but the horses themselves are very slow, and it is this that will, I fear, mar your chances. The other drivers know less than you do, but their horses are fleeter; therefore, my dear son, see if you cannot hit upon some artifice whereby you may insure that the prize shall not slip through your fingers. The woodman does more by skill than by brute force; by skill the pilot guides his storm-tossed barque over the sea, and so by skill one driver can beat another. If a man go wide in rounding this way and that, whereas a man who knows what he is doing may have worse horses, but he will keep them well in hand when he sees the doubling-post; he knows the precise moment at which to pull the rein, and keeps his eye well on the man in front of him. I will give you this certain token which cannot escape your notice. There is a stump of a dead tree – oak or pine as it may be – some six feet above the ground, and not yet rotted away by rain; it stands at the fork of the road; it has two white stones set one on each side, and there is a clear course all round it. It may have been a monument to some one long since dead, or it may have been used as a doubling-post in days gone by; now, however, it has been fixed on by Achilles as the mark round which the chariots shall turn; hug it as close as you can, but as you stand in your chariot lean over a little to the left; urge on your right-hand horse with voice and lash, and give him a loose rein, but let the left-hand horse keep so close in, that the nave of your wheel shall almost graze the post; but mind the stone, or you will wound your horses and break your chariot in pieces, which would be sport for others but confusion for yourself. Therefore, my dear son, mind well what you are about, for if you can be first to round the post there is no chance of any one giving you the goby later, not even though you had

Adrastus's horse Arion[2] behind you horse which is of divine race – or those of Laomedon, which are the noblest in this country.

(Il. 23.306–372)

9.3 On Curvilinear Motion

According to Newton's First Law of motion, in the absence of forces acting on a rigid body,[3] it either rests or it moves at constant velocity, i.e., at constant speed and in a constant direction.[4] Velocity is a *vectorial entity*, i.e., to be defined both its meter (speed) and its direction are required. For example, the speed of a car can be read on its speedometer, but its direction is controlled by the driver through the wheel. Accordingly, constant velocity means constant both speed and direction, i.e., uniform motion along a straight line. This constant velocity can only be changed by a force acting on the body. This effect is governed by Newton's Second Law according to the simple equation:

$$\bar{F} = m \cdot \bar{a}$$

where $\bar{F}$ is force, m mass and $\bar{a}$ acceleration, i.e., force is equal to velocity rate times mass.[5] The following conclusions can be drawn:

If, at a certain moment, a force acts on the moving body in or against its direction of motion, the body is *accelerated or decelerated*, i.e., its speed in the direction of motion increases or decreases respectively. But if the force acts laterally or at an angle to the direction of motion, the direction changes as well.

To remain on a *curvilinear course* (Figure 9.6), a body must be acted upon by a lateral force, which, if removed, causes the body to leave its course and move in the tangential direction. With a circular course, the lateral force with constant meter F_n is constantly directed towards the centre (Figure 9.7).

This is the *centripetal* force equal to:

---

[2] This horse, Arion of Adrastus, was one of the favorite characters of the Thebaean Circle, still, outside the mythological Homeric circle. He was endowed with speech and reason and was related to the Arcadian cult of Poseidon (Neptune) and Demeter. In Pausanias (*Arcadian*, 25, 5), there are references to Thebais and Antimachus. According to tradition, this mythical horse has its origin in Poseidon. It is not certain how it came to Adrastus's possession.

[3] In fact, on a *particle*, i.e a material entity with no dimensions but with a mass equal to the mass of the solid body.

[4] S.A. Paipetis, *Engineering Mechanics*, Vol. I Statics, Ion Publishers, Athens 2003.

[5] Force and acceleration are vectorial entities, i.e., they can be defined by both their meter *and* their direction.

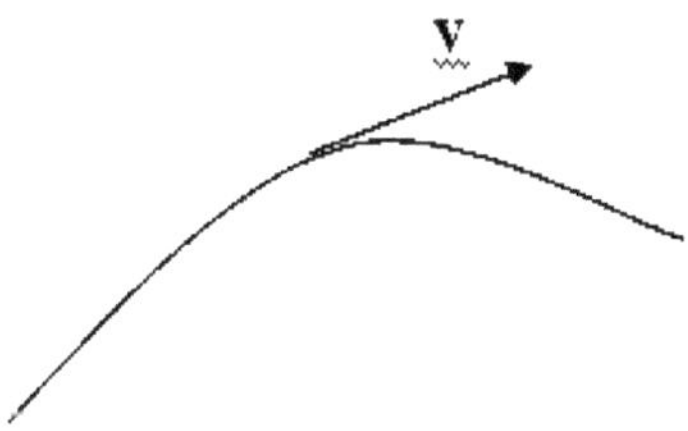

Figure 9.6 Motion on a curvilinear course.

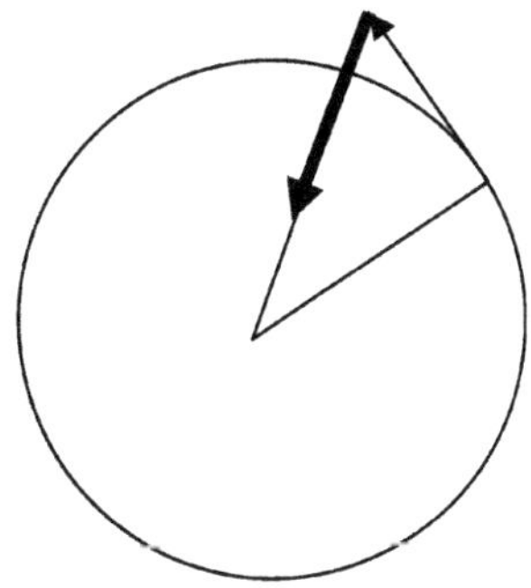

Figure 9.7 Motion on a circular course.

$$F_n = \frac{v^2}{R}$$

where v is speed of the body and R the radius of the circular course. This formula leads to the following remarks:

(a) With increasing speed, the centripetal force increases by the square of the increase: For example, if driving on a circular road, increasing speed by 10% causes the centripetal force to increases by 12.1%, while with increasing speed by 15%, the centripetal force increases by 32.25%.
(b) The increase of the latter, which is necessary for the vehicle to remain on the circular course, is felt by the driver as a force acting outwards, tending to divert the vehicle out of its course. This is the *centrigugal* force, not existing in reality, but simply expressing the tendency of the material body to move along the tangent of its circular course (or to resist any change of its kinematic condition).
(c) Decreasing radius of curvature R leads to increasing centripetal force. In fact, with a very small radius of curvature, i.e with very abrupt turns, the vehicle can hardly remain on course and, if friction is insufficient, it may jump out of the road.

This problem is especially important for air fighters engaged in midair combat, known colloquially as dog fighting: A plane with an enemy plane at its tail, must execute a U-turn as quickly as possible, to get at the tail of the enemy and use its weapons effectively. In this case, the speed increases abruptly, while the radius of the course decreases: The centripetal force imposes a great acceleration towards the centre, which may assume values of the order of nine g's or nine times the acceleration of gravity, e.g., the weight of the pilot increases by a ninefold.

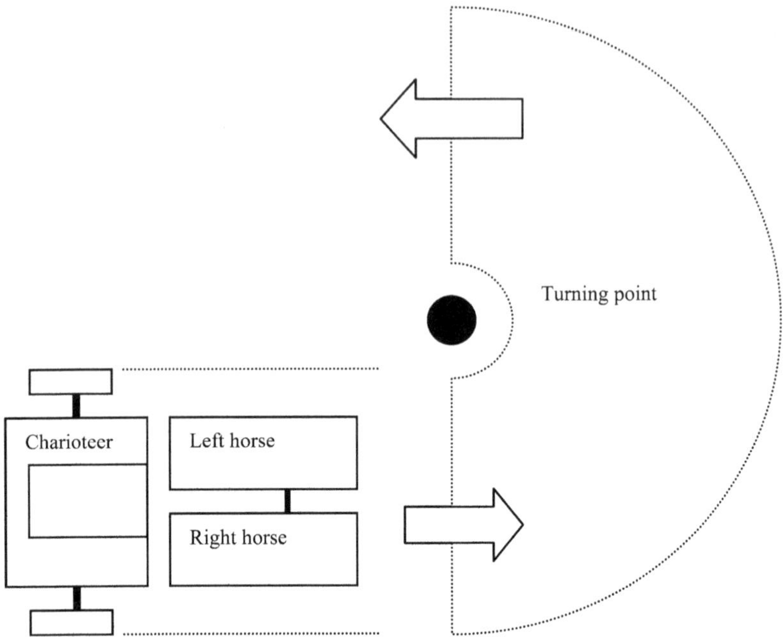

Figure 9.8 U-turn of the chariot.

9.4 The Chariot Race

According to the Homeric account, the race track had an oblong shape. The chariots started from one end towards the other, where they had to execute a quick 180-degree *anticlockwise* turn about a wooden pole and then return to the starting point. During this inversion of motion the chariot was moving along a semi-circular course with very small radius, and obviously this was the most difficult part of the race, by which not only should the horses be very fast, but also the charioteer should exhibit exceptional skill or he would not manage to control the inertial forces successfully. The following must be taken into account (Figure 9.8):

1. According to Nestor, Antilochus should move as close to the wooden pole as possible, but not too close, or the wheels might hit on one of the two rocks keeping the pole standing, and the chariot would be overturned. This is a proper choice, since the length of the semi-circular course S is proportional to its radius R, i.e., $S = \pi \cdot R$, where $\pi = 3.1416$, in other words, with double as much radius the chariot would

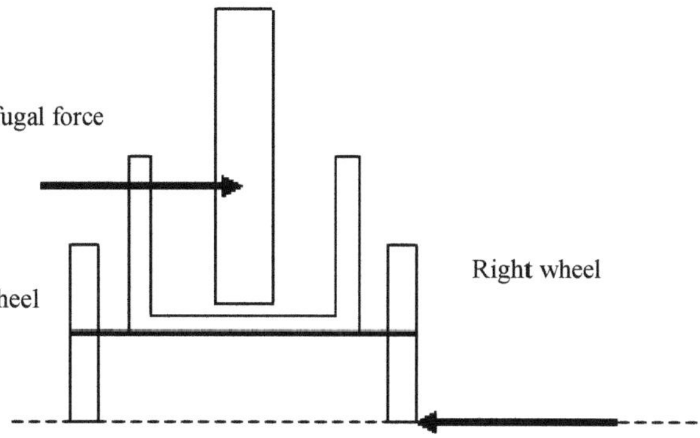

Figure 9.9 The couple of forces tending to overturn a vehicle on a curved course.

have to run along double as much distance. This choice corresponds to the well-known "internal course" in stadiums.

2. However, in this case, the radius of curvature for the horse on the left would be very small, and the centripetal force would increase too much, unless the speed of this horse would be reduced, in which case the ratio v^2/R remains within acceptable limits.
3. Things are different for the horse on the right, whose distance from the central point is greater than that of the left horse. Therefore, to keep both horses on the same radius, i.e., not to be detached from the yoke, the right horse *must accelerate*. According to Nestor, this is what Antilochus should manage, *by scream and lashes and by letting lose the horse's reins.*
4. The U-turn must be executed in the shortest time possible, i.e., with maximum speed but with the centripetal force not exceeding a limit, beyond which the chariot would overturned outwards. In fact, the centripetal force is applied on the chariot by friction at the contact point of the wheels with the ground, and is directed sidewise, i.e., towards the centre of the course. On the contrary, the inertial force, expressing the resistance of the chariot to a change of direction is manifested as the "centrifugal force", applied at the gravity centre of the chariot (of the system "chariot/charioteer", to be precise), and is directed outwards. These two forces constitute a couple corresponding to the *overturn moment* for the chariot (Figure 9.9). This can decrease (a) if the centre of gravity of the chariot moved lower and (b) if the charioteer, by moving his body to the left and

lower, created an opposite moment. This is exactly Nestor's advice that Antilochus should *bend the elastic chariot to the left.*

The difficulty to execute the U-turn, i.e., in minimizing inversion time successfully, is reflected in Nestor's statement that if Antilochus would achieve it first, *no other chariot, even with very fast horses, would be able to overcome*, which is in full agreement with the laws of curvilinear motion.

Chapter 10
Creep in Wood

> Homer knew that the first thing to do on getting your chariot out was to put the wheels on.[1]
>
> John Chadwick

In the Odyssey 4.39–42, Telemachus, Odysseus's son, travels to Sparta in search of information about his father from king Menelaus. Telemachus traveled in his chariot, which, upon arrival, the servants leaned against the wall. This event is accounted for within a single verse (*Od.* 4.42):

> ὁ δ' ἵππους μὲν λῦσαν ὑπὸ ζυγού ἱδρώοντας·
> καὶ τοὺς μὲν κατέδυσαν εφ' ἱππείησι κάπησι,
> πὰρ δ' ἔβαλον ζειάς, ἀνὰ δὲ κρῖ λευκὸν ἔμιξαν,
> **ἅρματα δ' ἔκλιναν πρὸς ἐνώπια ταμφανόωντα.**

> They took their sweating hands from under the yoke, made them fast to the mangers, and gave them a feed of oats and barley mixed. Then **they leaned the chariot against the end wall** of the courtyard.

It was necessary to position the chariot in this specific way, or even to remove the wheels when out of service, to keep the latter in good condition. Mythology tells of goddess Hebe, daughter of Zeus and Hera, the maid servant of the gods, whose duty was to serve them with ambrosia and nectar and, as additional duty, to put the wheels of the chariot of blue-eyed Athena in place every morning. If the wheels remained still under load by the chariot's own weight for long time, they would deform loosing their circular shape. Once this happened, it was impossible to restore them for immediate use, and the charioteer would find himself in an extremely unpleasant condition.

[1] *The Decipherment of Linear B*, Cambridge University Press, 1968.

S.A. Paipetis, *The Unknown Technology in Homer*, History of Mechanism and Machine Science 9,
DOI 10.1007/978-90-481-2514-2_10,

The development of chariots of various kinds, as well as reconstructions of the Mycenaean chariot are shown in Chapter 9 (Figures 9.1 through 9.3). In the present chapter, the property of light flexible wheels of Mycenaean chariots to deform under static load with time is investigated.[2]

The problem is not critical with chariot wheels of later times, although it is present with them too.[3] The deformation of bows and chariot wheels under prolonged loading is termed as "creep". In elementary elasticity, it is assumed that materials loaded are capable of carry this load *ad infinitum*, undergoing a deformation upon application of the load, which remains constant and invariable thereafter, e.g., it would not change with time, if load remains constant. However, things are different in practice: almost all materials, when subjected to constant load, keep deforming, or creeping, with time.

The creep rate varies for different materials. For example, wood, ropes and concrete exhibit creep, which must be taken into account in the design of structures. The same occurs with fabrics: clothes loose their shape, leggings turn baggy at the knees, etc., especially wool and cotton fabrics as well as those made of synthetic fibres.

In general, metals creep much less than non-metals, for example, steel creeps substantially under heavy load and at high temperature, however, with steel structures operating at ambient temperature, creep is negligible.

Creep often causes redistribution of mechanical stresses in structures, since at heavily loaded positions they creep faster. As a result, old shoes are generally more comfortable than new: creep adjusts their shape to the foot shape.

The effect of creep is evident with old wooden structures, such as wooden roofs or even boats, whose ends keep sinking, while the central part is rising. It is also known, that steel car suspension springs are receding with time and must be replaced.

Finally, although creep rate varies with various materials, all materials creep more or less in a similar way. For example, in Figure 10.1, the variation of strain for a certain material is plotted against time (in fact, logarithmic time, to shorten time scale) and for various load levels. Note that, under a critical load, a creeping material never breaks, while, under higher loads, it is a matter of time until break occurs.

[2] J.E. Gordon, *Structures*, Penguin Books Ltd., Harmondsworth, Middlesex, UK, 1978, pp. 146.

[3] Professor Gordon tells of stories about V.I.P.'s getting seasick when riding in state-coaches.

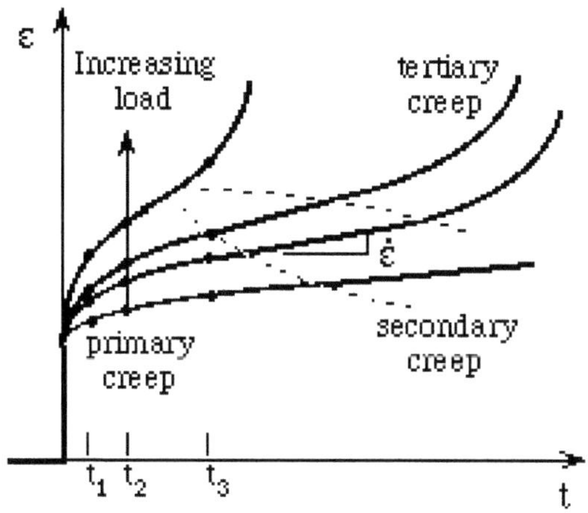

Figure 10.1 Typical creep behaviour: strain ε vs. time t, at different load levels.

The Homeric account along with existing representations of Mycenaean chariots shows that the Greeks of the time were well aware of the mechanical properties of structural materials and knew how to use this knowledge effectively.

Chapter 11
Hydrodynamics of Vortices and the Gravitational Sling

In Book XII of the Odyssey, Odysseus tells of his passing through the straits of Scylla and Charybdis on the instructions of Circe the sorceress (Figures 11.1–11.3), and explains how he could sail through the horrendous straits safely:

> Οἱ δὲ δύω σκόπελοι ὁ μὲν οὐρανὸν εὐρύν ἱκάνει
> ὀξείῃ κορυφῇ, νεφέλη δὲ μιν ἀμφιβέβηκε
> κυανέη τὸ μὲν οὔποτ' ἐρωεῖ, οὐδὲ ποτ' αἴθρη
> κείνου ἔχει κορυφὴν οὔτ' ἐν θέρει οὔτ' ἐν ὀπώρῃ
> οὐδὲ κεν ἀμβαίη βροτὸς ἀνήρ, οὐδ' ἐπιβαίη,
> οὐδ' εἴ οἱ χεῖρές τὲ ἐείκοσι καὶ πόδες εἶεν·
> πέτρη γὰρ λίς ἐστι, περιξεστῇ ἐϊκυῖα.
> μέσσῳ δέ ἐν σκοπέλῳ ἐστὶ σπέος ἠεροειδές,
> πρὸς ζέποη εἰς 'Ερεβος τετραμμένον, ᾗ πὲρ ἄν ὑμεῖς
> νῆα παρὰ γλαφυρὴν ἰθύνετε, φαίδιμ' 'Οδυσσεῦ.
> οὐδὲ κεν ἐκ νηὸς γλαφυρῆς αἰζήϊος ἀνὴρ
> τόξῳ ὀϊστεύσας κοῖλον σπέος εἰσαφίκοιτο.
> ἔνθα δ' ἐνί Σκύλλη ναίει δεινὸν λελακυῖα.
> τῆς ἤ τοι φωνὴ μὲν ὅση σκύλακος νεογιλλῆς
> γίγνεται, αὐτὴ δ' αὖτε πέλωρ κακόν· οὐδὲ κὲ τὶς μιν
> γηθήσειεν ἰδών, οὐδ' εἰ θεὸς ἀντιάσειεν.
> τῆς ἤ τοι πόδες εἰσὶ δυώδεκα πάντες ἄωροι,
> ἐξ δὲ τὲ οἱ δειραὶ περιμήκεες, ἐν δὲ ἑκάστῃ
> σμερδαλέη κεφαλή, ἐν δὲ τρίστοιχοι ὀδόντες,
> πυκνοὶ καὶ θαμέες, πλεῖοι μέλανος θανάτοιο.
> μέσση μὲν τὲ κατὰ σπείους κοίλοιο δέδυκεν
> ἔξω δ' ἔξίσχει κεφάλας δεινοῖο βερέθρου·
> αὐτοῦ δ' ἰχθυάᾳ, σκόπελον περιμαιμώωσα,
> δελφῖνάς τε κύνας τε καὶ εἴ ποθι μεῖζον ἕλῃσι
> κῆτος, ἅ μυρία βόσκει ἀγάστονος 'Αμφιτρίτη.
> τῇ δ' οὔ πώ ποτὲ ναῦται ἀκήριοι εὐχετόωνται

S.A. Paipetis, *The Unknown Technology in Homer*, History of Mechanism and Machine Science 9, DOI 10.1007/978-90-481-2514-2_11, © Springer Science + Business Media B.V. 2010

παρφυγέειν σὺν νηί· φέρει δὲ τε κρατὶ ἑκάστῳ
φῶτ' ἐξαρπάξασα νεός κυανοπρῴροιο.
Τὸν δ' ἕτερον σκόπελον χθαμαλώτερον ὄψει, 'Οδυσσεῦ,
πλησίον ἀλλήλων· καὶ κεν διοϊστεύσειας.
τῷ δ' ἐν ἐρινεὸς ἔστι μέγας, φύλλοισι τεθηλῶς·
τῷ δ' ὑπὸ δῖα Χάρυβδις ἀναρρυβδεῖ μέλαν ὕδωρ.
τρὶς μὲν γὰρ τ' ἀνίησιν ἐπ' ἤματι, τρὶς δ' ἀναροιβδεὶ
δεινόν· μὴ σὺ γὲ κεῖθι τύχοις, ὅτι ροιβδήσειεν·
οὐ γὰρ κεν ρύσαιτο σ' ὑπὲκ κακοῦ οὐδ' ἐνοσίχθων.
ἀλλὰ μάλα Σκύλλης σκοπέλω πεπλημένος ὦκα
νῆα πάρεξ ἐλάαν, ἐπεὶ ἦ πολὺ φέρτερον ἐστιν
ἓξ ἑτάρους ἐν νηὶ ποθήμεναι ἢ ἅμα πάντας.

Of these two rocks the one reaches heaven and its peak is lost in a dark cloud. This never leaves it, so that the top is never clear not even in summer and early autumn. No man though he had twenty hands and twenty feet could get a foothold on it and climb it, for it runs sheer up, as smooth as though it had been polished. In the middle of it there is a large cavern, looking West and turned towards Erebus; you must take your ship this way, but the cave is so high up that not even the stoutest archer could send an arrow into it. Inside it Scylla sits and yelps with a voice that you might take to be that of a young hound, but in truth she is a dreadful monster and no one – not even a god – could face her without being terror-struck. She has twelve mis-shapen feet, and six necks of the most prodigious length; and at the end of each neck she has a frightful head with three rows of teeth in each, all set very close together, so that they would crunch any one to death in a moment, and she sits deep within her shady cell thrusting out her heads and peering all round the rock, fishing for dolphins or dogfish or any larger monster that she can catch, of the thousands with which Amphitrite teems. No ship ever yet got past her without losing some men, for she shoots out all her heads at once, and carries off a man in each mouth.
You will find the other rocks lie lower, but they are so close together that there is not more than a bowshot between them. [A large fig tree in full leaf grows upon it], and under it lies the sucking whirlpool of Charybdis. Three times in the day does she vomit forth her waters, and three times she sucks them down again; see that you be not there when she is sucking, for if you are, Neptune himself could not save you; you must hug the Scylla side and drive ship by as fast as you can, for you had better lose six men than your whole crew.

(Od. 12.73–110)

Scylla was a mythical monster, originating from the Mycenaean or perhaps the Minoan era, appearing under different forms and ways of action. According to Evans (B8A IX, p. 58) the dog-like head of Scylla has its origin in the Minoan civilization. Nilson identifies the Minoan dog-like head of Scylla with the Scylla of Odyssey (Nilson, *Minoan-Mycenaean Religion*).

Figure 11.1 Giovanni Benedetto Castiglione (Il Grechetto), 1610–1665: Circe With Companions of Ulysses Changed Into Animals, ca. 1655, Fine Arts Museum of San Francisco (reproduced bu permission).

S. Marinatos objects both views (Minoan and Homeric Scylla, A.D. pp. 51–62), maintaining that the Homeric Scylla is irrelevant to Minoan patterns, but probably has its origin in Mycenaean myths:

> Scylla, daughter of Forkys and Hecate, holding by her sides little dogs, is the master of the straits around Sicily. She is of monstrous size, with twelve legs, six heads and in every mouth three rows of teeth and eyes emitting fire. The rest of her body was hidden in a cave submerged in the sea. She was attached on the rock and her heads were hanging out, to reach the passing boats from the rock.

For the second monster, Charybdis, Homer does not make clear whether it was a beast or a natural phenomenon. However, the periodicity of water flow shows that this was a tidal vortex, due to the form of the coasts of the straits and of the sea bottom at that particular place between Adriatic and Tyrrenian Seas (Messina straits). The reference to Charybdis in the Odyssey is not the first. For example, in Orphica Argonautica an account is given in Argo, Jason's ship, managing to pass by Charybdis safely:

> Τῆμος δ' ἀντολίαισιν ἐγείρετο φωσφόρος αἴγλη,
> ὄρθριοι εἰρεσιῃς γλαυκὴν ἐχαράξαμεν ἅλμην,
> Σαρδῶον δ' ἰκόμεσθα βυθόν, κόλπους τε Λατίνων
> νήσους τ' Αὐσονίας, Τυρρηνάς δ' ἰόκομεθ' ἀκτᾶς.
> Αὐτάρ ѐπεί Λιλύβαιον ἐπέσχομεν ἠχέτα πορθμόν,

Figure 11.2 Lorenzo Gabrieri (1588–1654): Circe, the sorceress.

Τριγλώχινα τὲ νῆσον ἐπέσχομεν, 'Εγκελάδοιο
Αἰτναίη φλὸξ σφιν ἄρ' ἐρήτυεν μεμαῶτας,
δὴ τόθ' ὑπὲρ πρώρης ὀλὸν περιέζεεν ὕδωρ
νειόθεν (ἐκ μυχάτου δὲ βυθοῦ ῥοίβδησε Χάρυβδις
κύματι καχλάζοντι, καὶ ἱστίον ἄκρον ἴκανε).
Νῆα δ' ἄρ' αὐτόθι οἱ κατέχεν ῥόος, οὐδὲ μιν εἴα
προπροθέειν, οὐδ' αὐτίκ' ἀναρρώσεσθαι ὀπίσσω,
κοίλω ἐπὶ λυγρῷ δὲ περιστροφάδην ἀλάλητο.
"Η τάχα καὶ δύσεσθ' 'Αργὼ κατὰ βένθε' ἔμελλεν
εἰ μὴ πρεσβίστη θυγάτηρ ἁλίοιο γέροντος
εὐρυβίην Πηλῆα πόσιν λελίντρο ἰδέσθαι,
μειλιχίη δ' ἐκδῦτο βυθοῦ καὶ ῥύσσατ' ὀλέθρου
'Αργῴαν ἄκατον, καὶ ὑπ' ἰλύος ἐξεσάωσε.

... When, by sunrise, the light-bringing Dawn rose, since early morning we cut the blue sea by the oars, and reached Sardoon Sea and bays of the Latins and the islands of Ausonia[1] and the Tyrrenean coasts. And having approached the heavy-sounding channel, Lilybaeus, and got near the three-tongued island, the Aetneaean flame of Engeladus delayed us, although we were in a hurry and then murderous water from the bottom was boiling

[1] Ausonia, Italy.

Figure 11.3 Hubert Maurer (1738–1818): Odysseus and Circe, Gemäldegalerie der Akademie der bildende Künste (reproduced by permission).

> all around, covering the bow (from the depths of sea bottom Charybdis was spitting up with boiling waves) and reached the top of the sail. was holding the ship back and would not let it advance neither retreat, but it was cyclically turning around in the perilous cavity. Truly, Argo was going soon to submerge in the depths, if it were not for the older daughter of the seaborn old man,[2] who was desirous to see her most potent husband, Peleus, and she gently emerged from the sea bottom to rescue the boat, Argo, from catastrophe and deliver her from the muddy bottom. ...
>
> (Orphica Argonautica 1246–1263)

Further accounts originate from Apollodorus:[3]

> Μετὰ δὲ τάς Σειρήνας τὴν ναῦν Χάρυβδις ἐξεδέχετο καὶ
> Σκύλλα καὶ πέτραι πλαγκταί, υπὲρ ὧν φλὸξ πολλὴ καὶ

[2] He refers to Thetis, daughter of Nereus and mother to Achilles.

[3] Besides the above, there are numerous references to Charybdis, as in Virgil's Aeneid, III 418, 555, Strabo's *Geography*, VI 268 and finally in *Navigations d'Ulysse*, Vol. IV, pp. 390–405.

καπνὸς ἀναφερόμενος ἑωράτο. ἀλλὰ διὰ τούτων διεκόμισεν τὴν ναῦν σὺν Νηρηίσι Τέτις παρακληθεῖσα ὑπὸ Ἥρας.

After the Sirenes, the ship ended up to Scylla and Charybdis and the moving rocks,[4] over which lots of fire and smoke to emerge. But Thetis, on Hera's request, had the ship pass through them, assisted by the Nereids.

(Library A, ix, 25)

Μετὰ δὲ τοῦτο παραγίνεται ἐπὶ δισσάς όδους. ἔνθεν μὲν ἦσαν αἱ πλαγκταὶ πέτραι, ἔνθεν δὲ ὑπερμεγέθεις σκόπελοι δύο. ἦν δὲ ἐν μὲν θατέρῳ Σκύλλα, Κραταιίδος θυγάτηρ καὶ Τριήνου ἢ Πόρκου, πρόσωπον ἔχουσα καὶ στέρνα γυναικός, ἐκ λαγόνων δὲ κεφαλὰς ἕξ καὶ δώδεκα πόδας κυνῶν. ἐν δὲ θατέρῳ [τῷ σκοπέλῳ] ἦν χάρυβδις, ἥ τῆς ἡμέρας τρὶς ἀνασπῶσα τὸ ὕδωρ πάλιν ἀϝίει. ὑποθεμένης δὲ Κίρκης, τὸν μὲν παρὰ τὰς Πλαγκτᾶς πλοῦν ἐφυλάξατο, παρὰ δὲ τὸν τῆς Σκύλλης σκόπελον [παραπλέων] ἐπὶ τῆς πρύμνης ἔστη καθωπλισμένος. ἐπιφανεῖσα δὲ ἡ Σκύλλα ἕξ ἑταιρους ἁρπάσασα τούτους κατεβίβρωσκεν. λυθείσης δὲ τῆς νεὼς 'Οδυσσεὺς τὸν ἱστὸν κατασξὼς παραγίνεται εἰς τὴν Χάρυβδιν. Τῆς δὲ Χαρύβδεως καταπινούσης τὸν ἱστόν, ἐπιλαβόμενος ὑπερπεφυκότος ἐρινεοῦ περιέμεινε. Καὶ πάλιν ἀνεθέντα τὸν ἱστὸν θεωρήσας, ἐπὶ τοῦτον πίφας εἰς 'Ογυγίαν νῆσον διεκομίσθη.

And in the other cliff was Charybdis, who thrice a day drew up the water and spouted it again. By the advice of Circe he shunned the passage by the Wandering Rocks, and in sailing past the cliff of Scylla he stood fully armed on the poop. But Scylla appeared, snatched six of his comrades, and gobbled them up. And thence he came to Thrinacia, an island of the Sun, where kine were grazing, and being windbound, he tarried there. But when his comrades slaughtered some of the kine and banqueted on them, for lack of food, the Sun reported it to Zeus, and when Ulysses put out to sea, Zeus struck him with a thunderbolt. And when the ship broke up, Ulysses clung to the mast and drifted to Charybdis. And when Charybdis sucked down the mast, he clutched an overhanging wild fig-tree and waited, and when he saw the mast shot up again, he cast himself on it, and was carried across to the island of Ogygia.

(Epitome vii, 21 & 23)

[4] *Symplegades* or Blue Rocks were at the entrance to the Black Sea.

11.1 Hydrodynamics of Vortices[5]

The liquid surface of a vortex is mainly formed under the effect of two forces: gravity, attracting downwards the water particles, trying to keep its surface horizontal, and buoyancy, i.e., the Archimedean force acting upwards. Both of these forces eventually create (a) a centripetal component (and, respectively, a centripetal acceleration), maintaining the particle on a circular course, and (b) a tangential component, driving the particle along this course at constant speed. So the system is directed to a steady-state condition, by which the free surface of the water assumes the form of an inverted bell. In this condition, all of the forces acting on every particle have zero resultant, and, accordingly, every particle remains at the same place *ad infinitum*, unless a third force acts, which will upset equilibrium. At this particular particle position (defined by its rotation radius and, respectively, by its height within the vortex), there is a corresponding speed, which, if changed, equilibrium breaks down.

This is in fact the same behavior as that of a satellite rotating around a planet on a circular orbit, by which, at a specific height, a respective speed develops maintaining equilibrium. However, on a ship sailing within a vortex, more forces are acting, such as friction between its surfaces and the water, air resistance etc. Such forces tend to reduce its speed in relation to the speed required to set the ship on course around the vortex and the forces do not remain in equilibrium. Then, the ship, under the influence of gravity, moves helicoidally towards the centre of the vortex. The respective verses of Circe's instruction to Odysseus are here repeated:

> ... the sucking whirlpool of Charybdis. Three times in the day does she vomit forth her waters, and three times she sucks them down again; see that you be not there when she is sucking, for if you are, Neptune himself could not save you; you must hug the Scylla side and drive ship by as fast as you can, for you had better lose six men than your whole crew.
>
> (Od. 7.103–106)

Analogous is the effect of the satellite orbiting a planet, in the presence of air resistance. Besides surface overheating, the vehicle decelerates, and if out of control, his initially circular orbit turns into a steep helix, and, as a result, the same or its burnt remnants may drop on the surface of the planet.

---

[5] Analysis conducted by Professor George Ch. Vatistas, Concordia University, Canada: "Floating Body Dynamics Inside Whirlpools Found in Mythology and Literature", in *Proceedings of the 2nd International Congress on Ancient Greece and Modern World*, Ancient Olympia, 12–17 July 2002, University of Patras Publications.

The key lies in the expression "drive ship by as fast as you can", in other words "move fast, to account speed loss due to friction and remain in course instead of diving to the bottom". In the sequence, the Homeric account tells of the events, when Odysseus's boat approached the straits:

> ἀλλ' ὅτε δὴ τὴν νῆσον ἐλείπομεν, αὐτίκ' ἔπειτα
> καπνὸν καὶ μέγα κῦμα ἴδον καὶ δοῦπον ἄκουσα.
> τῶν δ' ἄρα δεισάντων ἐκ χειρῶν ἔπτατ' ἐρετμά,
> βόμβησαν δ' ἄρα πάντα κατὰ ῥόον· ἔσχετο δ' αὐτοῦ
> νηῦς, ἐπεὶ οὐκέτ' ἐρετμὰ προήκεα χερσὶν ἔπειγον.

> ... after we had got past the island I saw a great wave from which spray was rising, and I heard a loud roaring sound. The men were so frightened that they loosed hold of their oars, for the whole sea resounded with the rushing of the waters, but the ship stayed where it was, for the men had left off rowing.
>
> (Od. 7.201–205)

The word δούπος (= *roaring sound*), in this case, means the sound produced by a wave breaking on a rocky coast in the vicinity of uncovered underwater caves. The Homeric account also describes smoke or steam raising, which leads to the conclusion that an enormously high wave developed, with all tidal characteristics, which broke most powerfully on Scylla's coast, possessing many abrupt cliffs, creating thus Charybdis's phenomenon.

Homer provides clear remarks on the physical creation of the vortex in the following passage:

> ἡμεῖς μὲν στεινωπὸν ἀνεπλέομεν γοόωντες·
> ἔνθεν μὲν Σκύλλη, ἑτέρωθι δὲ δῖα Χάρυβδις
> δεινὸν ἀνερροίβδησε θαλάσσης ἁλμυρὸν ὕδωρ.
> ἦ τοι ὅτ' ἐξεμέσειε, λέβης ὣς ἐν πυρὶ πολλῷ
> πᾶσ' ἀναμορμύρεσκε κυκωμένη, ὑψόσε δ' ἄχνη
> ἄκροισι σκοπέλοισιν ἐπ' ἀμφοτέροισιν ἔπιπτεν·
> ἀλλ' ὅτ' ἀναβρόξειε θαλάσσης ἁλμυρὸν ὕδωρ,
> πᾶσ' ἔντοσθε φάνεσκε κυκωμένη, ἀμφὶ δὲ πέτρη
> δεινὸν ἐβεβρύχει, ὑπένερθε δὲ γαῖα φάνεσκε
> ψάμμῳ κυανέῃ· τοὺς δὲ χλωρὸν δέος ᾕρει.

> Then we entered the Straits in great fear of mind, for on the one hand was Scylla, and on the other dread Charybdis kept sucking up the salt water. As she vomited it up, it was like the water in a cauldron when it is boiling over upon a great fire, and the spray reached the top of the rocks on either side. When she began to suck again, we could see the water all inside whirling round and round, and it made a deafening sound as it broke against the rocks.

> We could see the bottom of the whirlpool all black with sand and mud, and the men were at their wit's ends for fear.
>
> (Od. 7.235–243)

Since the phenomenon is of dynamic nature, it evolves through various stages. The all-powerful tidal wave, in conjunction with the local morphological characteristics, creates a vortex. In the presence of the centripetal field, the water tends to assume the form of an inverted bell. The central lowering starts propagating towards the sea bottom, reaching a critical value, depending of the power of the whirlpool: With very powerful, fully developed whirlpools, the sea bottom may be uncovered. It is reasonable to assume that by the phrase "the water all inside whirling round and round, and it made a deafening sound", Homer describes the effect of unstable waves, which, as it is well-known, accompany vortices.[5]

Based on the above, one can distinguish three cases: (a) the boat in the vortex moves at the speed of the stream; then it remains on a constant course at constant height, (b) due to friction, it looses speed in relation to the stream, then the boat is pulled to the bottom by its own weight, and (c) with the hard efforts of the oarsmen, the boat speed overcomes the speed of the stream; then the boat manages to fly away from the stream and get out of the vortex.

So, it can be concluded, that the hydrodynamic attributes of the ship, the propulsion power available and the power of the vortex define a limit circle, above which the ship survives. The term "event horizon", used with neutron stars or black holes, believed to exist in the centers of the galaxies, finds here a plausible application.

If Charybdis phenomenon was in full development when Odysseus was about to cross the straits, he should apply a strategy ensuring the best chances of survival. If he were steering the boat along the coast opposite to Scylla, he would have met great resistance, having to move against a very strong stream directly and he would never survive. Therefore, he had to move along the stream and not against it, to maximize his speed. Moreover, he would have to sail at the maximum possible distance from the centre of the vortex, where the attraction towards the centre would be minimal. Therefore, the only solution was to move to the right, steering the ship as close to Scylla as possible, i.e., to "drive the ship by as fast as he could". And, to maximize the ship's propulsion power, gives the oarsmen and the coxswain the following orders:

---

[5] Thomson, W. (Lord Kelvin), Vibrations of a vortex column, *Philos. Mag.*, vol. 10, 1880, 155; and Vatistas, G.H., A note on liquid vortex sloshing and Kelvin's equilibria, *Journal of Fluid Mechanics*, vol. 217, 1990, 241–248.

ὑμεῖς μὲν κώπῃσιν ἁλὸς ῥηγμῖνα βαθεῖαν
τύπτετε κληΐδεσσιν ἐφήμενοι, αἴ κέ ποθι Ζεὺς
δώῃ τόνδε γ' ὄλεθρον ὑπεκφυγέειν καὶ ἀλύξαι·
σοὶ δέ, κυβερνῆθ', ὧδ' ἐπιστέλλομαι· ἀλλ' ἐνὶ θυμῷ
βάλλευ, ἐπεὶ νηὸς γλαφυρῆς οἰήια νωμᾷς·
τούτου μὲν καπνοῦ καὶ κύματος ἐκτὸς ἔεργε
νῆα συ δὲ σκοπέλον ἐπιμαίεο, μή σε λάθῃσι
κεῖσ' ἐξορμήσαντα καὶ ἐς κακὸν ἄμε βάλησθα.
ὣς ἐφάμην, οἱ δ' ὦκα ἐμοῖς ἐπέεσι πίθοντο.
Σκύλλην δ' οὐκέτ' ἐμυθεόμην, ἄπρηκτον ἀνίην,
μή πώς μοι δείσαντες ἀπολλήξειαν ἑταῖροι
εἰρεσίης, ἐντὸς δὲ πυκάζοιεν σφέας αὐτούς.

therefore, let us all do as I say, trust in Jove and row on with might and main. As for you, coxswain, these are your orders; attend to them, for the ship is in your hands; turn her head away from these steaming rapids and hug the rock, or she will give you the slip and be over yonder before you know where you are, and you will be the death of us.

(Od. 7.214–225)

As a general conclusion, the Greeks of the Homeric era seem to possess sufficiently deep knowledge of the forces creating vortices in the sea, i.e., gravity, centripetal force and buoyancy. In fact, there is substantial evidence that Greek sailors possessed at least qualitative knowledge of the main properties of vortices, such as surface lowering at the centre and their attractive or repulsive behavior, depending on the particular case, and used it skillfully, when encountered such dangerous phenomena during their travels.

11.2 The Gravitational Sling

A very interesting remark is that this fundamental knowledge, familiar to the ancient Greeks, is nowadays used in space technology, for the so-called *gravitational sling*, i.e., the use of the gravitational attraction of a planet, to change course and velocity of a space vehicle without energy consumption: It is a very common maneuvre applied with space vehicles moving towards the outer planets of the solar system, which, otherwise, might need years to accomplish.

Consider a space vehicle on an orbit leading it near Jupiter (Figure 11.4). By approaching Jupiter, it is attracted by its gravity and accelerates. After passing the planet, the gravity, still attracting the vehicle, causes it to decel-

Figure 11.4 The planet Jupiter is used for the gravitational sling, due to its great size and strong gravity (diameter at the equator 142,984 km, 11,209 times that of the Earth, mass 1,899 × 10^{27} kg, 317.8 times that of the Earth, acceleration of gravity at the equator 23.12 km/s^2, 2.358 times that of the Earth).

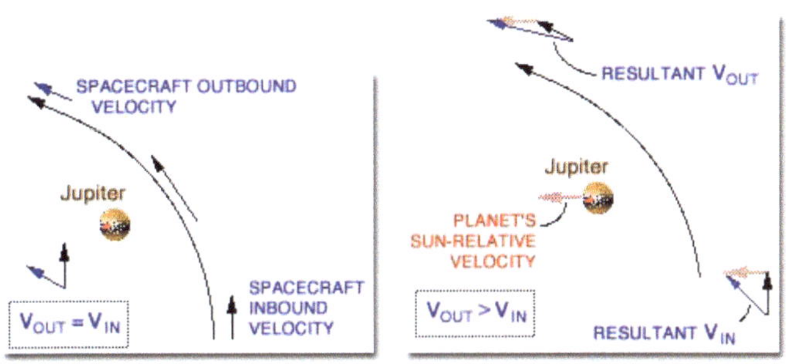

Figure 11.5 Gravitational sling of a space vehicle assisted by the attraction of Jupiter: (a) initial V_{IN} and final velocity V_{OUT} in relation to the planet and (b) the same in relation to the Sun.

erate. Speed remains the same, but the direction of motion has meanwhile changed.

However, the planets themselves do not remain motionless, but move along their orbits around the Sun. Therefore, although the speed of the vehicle remains unchanged in relation to Jupiter, initial and final velocities are different in relation to the Sun (Figure 11.5). According to the direction of the exit branch of the course, the space vehicle may gain substantial percentage of the course speed of the planet, which in the case of Jupiter, may exceed 13 km/sec!

Part 3
Automation and Artificial Intelligence

Chapter 12
The Forge of Hephaestus

In the ancient Greek religion, Hephaestus (Vulcan) was a god of fire. Initially, he was a deity of Asia Minor and the Aegean islands. He held an important position among the Olympian gods. Hephaestus was the son of Zeus and Hera and was born lame at both his legs. His mother could not bear his lameness and, after a serious argument with his father, hurled him from Olympus into the Aegean Sea, near Lemnus island. Thetis, a Nereid, who was to be the mother of Achilles, and Eurynome, her sister, saved him. His marriage to Aphrodite (Venus) was ill-chosen and ill-fated. As the god of fire, Hephaestus became the heavenly metallurgist and patron of metal workers. It was believed that fires emitted from volcanoes marked the position of his several forges (Figure 12.1). According to Mythology, he was fashioning the thunderbolts of Zeus. His worship reached Athens before 600 BCE. In artistic portrayals, Hephaestus is usually depicted as a middle-aged man with a beard, wearing a short, sleeveless robe and a round, tight cap, covering his untidy hair.

In the forge of Hephaestus, the most advanced machines of the times existed, as described by Homer, e.g., automated and autonomous machines, as follows:

(a) *The self-propelled tripods*
According to the Iliad, Book XVIII, Thetis, Achilles's mother, visited Hephaestus, begging of him to fashion an armor for her son. She found him engaged, building a number of wheeled tripods capable of moving by themselves into the meeting place of the gods and coming back by the end of the meeting. This account describes autonomous devices capable of performing simple repetitive work, such as serving food:

S.A. Paipetis, *The Unknown Technology in Homer*, History of Mechanism and Machine Science 9, DOI 10.1007/978-90-481-2514-2_12,

Figure 12.1 Giordano Luca (1634–1705): Vulcan's Forge, Oil on canvas, The State Hermitage Museum, St. Petersburg (photograph © The State Hermitage Museum, reproduced by permission).

ὥς οἱ μὲν τοιαῦτα πρὸς ἀλλήλους ἀγόρευον·
Ἡφαίστου δ' ἵκανε δόμον Θέτις ἀργυρόπεζα
ἄφθιτον ἀστερόεντα μεταπρεπέ ἀθανάτοισι·
χάλκεον, ὅν ῥ' αὐτὸς ποιήσατο κυλλοποδίων.
τὸν δ' εὗρ' ἱδρώοντα ἑλισσόμενον περὶ φύσας
σπεύδοντα· τρίποδας γὰρ ἐείκοσι πάντας ἔτευχεν
ἑστάμεναι περὶ τοῖχον ἐϋσταθέος μεγάροιο,
χρύσεα δὲ σφ' ὑπὸ κύκλα ἑκάστω πυθμένι θῆκεν,
ὄφρά οἱ αὐτόματοι θεῖον δυσαίατ' ἀγῶνα
ἠδ' αὖτις πρὸς δῶμα νεοίατο θαῦμα ἰδέσθαι.

> Thus did they converse. Meanwhile Thetis came to the house of Hephaestus, imperishable, star-bespangled, fairest of the abodes in heaven, a house of bronze wrought by the lame god's own hands. She found him busy with his bellows, sweating and hard at work, for he was making twenty tripods that were to stand by the wall of his house, and he set wheels of gold under them all that they might go of their own selves to the assemblies of the gods, and come back again- marvels indeed to see. They were finished all but the ears of cunning workmanship which yet remained to be fixed to them: these he was now fixing, and he was hammering at the rivets. While he was thus at work silver-footed Thetis came to the house.
>
> (Il. 18.368–377)

(b) *The automatic bellows*

Hephaestus promises Thetis that he will not be long and gets to work with his fire, equipped with a set of bellows[1] possessing all the features of a super-automatic device.

> ὣς εἰπὼν τὴν μὲν λίπεν αὐτοῦ, βῆ δ' ἐπὶ φύσας·
> τὰς δ' ἐς πῦρ ἔστρεψεν κέλευσαι τε ἐρψγάζεσθαι.
> φῦσαι δ' ἐν χοάνοισιν ἐείκοσι πᾶσαι ἐφύσων
> παντοίην εὔπρηστον ἀϋτμὴν ἐξανεῖσαι,
> ἄλλοτε μὲν σπεύδοντι παρέμμεναι, ἄλλοτε δ' αὖτε,
> ὅπως Ἥφαιστός τ' ἐθέλοι καὶ ἔργον ἄνοιτο.

> When he had so said he left her and went to his bellows, turning them towards the fire and bidding them do their office. Twenty bellows blew upon the melting-pots, and they blew blasts of every kind, some fierce to help him when he had need of them, and others less strong as Hephaestus willed it in the course of his work.
>
> (Il. 18.468–473)

Based on this account of the device which activates the bellows, one may assume the following:

1. The bellows operate through a natural energy source, since there is no reference to workers doing this job. Moreover, if this operation were executed by humans, then, they could easily be directed towards the fire, and no automatic devices would be necessary.
2. There is a mechanism directing the bellows toward the point that Hephaestus wants the fire steered.
3. There is a mechanism conceiving, interpreting and activating Hephaestus' wishes and directs the bellows accordingly. This appears as a fully

[1] N. Orfanoudakis, The Bellows Go to War, in *Extraordinary Machines and Structures in Antiquity*, S.A. Paipetis (Ed.), Peri Technon Publ., Patras, Greece, 2002.

autonomous system, since Hephaestus "ordered the bellows to work", while it is physically impossible the work to be controlled by Hephaestus himself, for example, with a foot-operated mechanism, since he was lame with his legs deformed, so that he had a great difficulty to walk without external assistance, afforded to him by "two golden girls", as it will be shown in the sequence.

(c) *Hephaestus' traps*

Hephaestus had no inhibitions in using his crafts to punish those who had wronged him. The first one to suffer the consequences of her deeds was Hera, his own mother, on whom he wanted to revenge for the way she treated him as an infant: He gave her a beautiful throne as a present, which captured and fettered her, when she tried to sit on it. Several writers *(Alcaeus 3, 49, Pindar 2, 83)* refer to a machine developed by Hephaestus to punish his mother. Even Pausanias *(Attics, 20, 3),* describing the southern side of the Acropolis of Athens, writes:

> The most ancient sanctuary of Dionysus lies beside the theatre. In the yard, there are two temples and two Dionysuses, one from Eleuterae and one made by Alcamenes of ivory and gold. Here, there are paintings also, one with Dionysus leading Hephaestus to the seat, on which Hera was tied down as soon as she sat on it. Hephaestus would not be convinced[2] to release her by any other god. Then Dionysus, who was greatly trusted by Hephaestus, made him drunk and led him to heavens. According to the tradition of the Greeks, Hera hurled new-born Dionysus away, and he, growing malicious, gave her as a gift a golden throne with invisible fetters, which tied Hera down as soon as she sat on it. All of these are painted, as well as Pentheus and Lycurgus, who are punished for their hybris to the god, and Ariadne, who is sleeping while Theseus is sailing away from her and Dionysus, who comes to snatch her and take her to heavens.

In the same way, Hephaestus treated his spouse Aphrodite or Charis, goddess of love and beauty. Aphrodite had an affair with Ares (Mars), god of war, and Helios rushed to inform Hephaestus (Figure 12.2), who punished the unfaithful lovers, while making love on his own bed, by means of a device that caught and tied them down (Figures 12.3–12.5). The Homeric account follows:

> Αὐτὰρ ὁ φορμίζων ἀνεβάλλετο καλὸν ἀείδειν
> ἀμφ' Ἄρεος φιλότητος ἐϋστεφάνου τ' Ἀφροδίτης,
> ὡς τὰ πρῶτα μίγησεν ἐν Ἡφαίστοιο δόμοισι
> λάθρῃ· πολλὰ δ' ἔδωκε, λέχος δ' ᾔσχυνεν καὶ εὐνὴν

[2] In order to return to Olympus and free Hera from her invisible fetters.

Figure 12.2 Diego Velázquez (1599–1660): La fragua de Vulcano, 1630, Prado Museum, Madrid: Helios informing Hephaestus of Aphrodite's unfaithfullness (reproduced by permission).

Ἡφαίστοιο ἄνακτος. ἄφαρ δέ οἱ ἄγγελος ἦλθεν
Ἥλιος, ὅ σφ' ἐνόησε μιγαζομένους φιλότητι.
Ἥφαιστος δ' ὡς οὖν θυμαλγέα μύθον ἄκουσε,
βῆ ῥ' ἴμεν ἔς χαλκεῶνα, κακὰ φρεσὶ βυσσοδομεύων·
ἐν δ' ἔθετ' ἀκμοθέτῳ μέγαν ἄκμονα, κόπτε δὲ δεσμοὺς
ἀρρήκτους ἀλύτους, ὄφρ' ἔμπεδον αὖθι μένοιεν.
αὐτὰρ ἐπεὶ δὴ τεῦξε δόλον κεχολωμένος Ἄρει,
βῆ ῥ' ἴμεν ἐς θάλαμον, ὅθι οἱ φίλα δέμνι' ἔκειτο·
ἀμφὶ δ' ἄρ' ἀρμίσιν χέε δέσματα κύκλῳ ἁπάντῃ,
πολλὰ δὲ καὶ καθύπερθε μελαθρόφιν ἐξεκέχυντο,
ἠΰτ' ἀράχνια λεπτά· τὰ γ' οὔ κὲ τις ἴδοιτο,
οὐδὲ θεῶν μακάρων· περὶ γὰρ δολόεντα τέτυκτο.
αὐτὰρ ἐπεὶ δὴ πάντα δόλον περὶ δέμνια χεῦεν,
ἔσατ' ἴμεν ἐς Λῆμνον, ἐΰκτίμενον πτολίεθρον,
ἥ οἱ γαιάων πολὺ φιλτάτη ἐστὶν ἁπασέων.
οὐδ' ἀλαοσκοπιὴν εἶχε χρησήνιος Ἄρης,
ὡς ἴδεν Ἥφαιστον κλυτοτέχνην νόσφι κιόντα·
βῆ δ' ἴμεναι πρὸς δῶμα περικλυτοῦ Ἡφαίστοιο,

ἰχανόων φιλότητος ἐϋστεφάνου Κυθερέης·
ἡ δὲ νέον παρὰ πατρὸς ἐρισθενέος Κρονίωνος
ἐρχομένη κατ' ἄρ' ἕζεθ' ὁ δ' εἴσω δώματος ἤει
ἐν τ' ἄρα οἱ φῦ χειρὶ ἔτος τ' ἔφατ' ἐκ τ' ὀνόμαζε·
"Δεῦρο, φίλη, λέκτρονδε, τραπείομεν εὐνηθέντες·
οὐ γὰρ ἔθ' Ἥφαιστος μεταδήμιος, ἀλλὰ που ἤδη
οἴχεται ἐς Λῆμνον μετὰ Σίντιας ἀγριοφώνους."
Ὣς φάτο, τῇ δ' ἀσπαστὸν ἐέσατο κοιμηθῆναι.
τὼ δε· ἐς δέμνια βάντε κατέδραμον ἀμφὶ δὲ δεσμοὶ
τεχνήεντες ἔχυντο πολύφρονος Ἡφαίστοιο,
οὐδὲ τι κινῆσαι μελέων ἦν οὐδ' ἀναεῖρε.
ὥς φάτο, τῇ δ' ἀσπαστὸν ἐέσατο κοιμηθῆναι.
τὼ δ' ἐς δέμνια βάντε κατέδραθον· ἀμφὶ δὲ δεσμοὶ
τεχνήεντες ἔχυντο πολύφρονος Ἡφαίστοιο,
οὐδέ τι κινῆσαι μελέων ἦν οὐδ' ἀναεῖραι.
καὶ τότε δὴ γίγνωσκον, ὅ τ' οὐκέτι φυκτὰ πέλοντο
ἀγχίμολον δέ σφ' ἦλθε περικλυτὸς ἀμφιγυήεις,
αὖτις ὑποστρέψας πρὶν Λήμνου γαῖαν ἱκέσθαι·
Ἠέλιος γάρ οἱ σκοπιὴν ἔχεν εἰπέ τε μῦθον.
βῆ δ' ἴμεναι πρὸς δῶμα φίλον τετιημένος ἦτορ·
ἔστη δ' ἐν προθύροισι, χόλος δέ μιν ἄγριος ᾕρει·
σμερδαλέον δ' ἐβόησε, γέγωνέ τε πᾶσι θεοῖσιν·
"Ζεῦ πάτερ ἠδ' ἄλλοι μάκαρες θεοὶ ἀεν ἐόντες,
δεῦθ', ἵνα ἔργα γελαστὰ καὶ οὐκ ἐπιεικτὰ ἴδησθε,
ὡς ἐμὲ χωλὸν ἐόντα Διὸς θυγάτηρ Ἀφροδίτη
αἰὲν ἀτιμάζει, φιλέει δ' ἀίδηλον Ἄρηα,
οὕνεχ' ὁ μὲν καλός τε καὶ ἀρτίπος, αὐτὰρ ἐγώ γε
ἠπεδανὸς γενόμην. ἀτὰρ οὔ τί μοι αἴτιος ἄλλος,
ἀλλὰ τοκῆε δύω, τὼ μὴ γείνασθαι ὄφελλον.
ἀλλ' ὄψεσθ', ἵνα τώ γε καθεύδετον ἐν φιλότητι
εἰς ἐμὰ δέμνια βάντες, ἐγὼ δ' ὁρόων ἀκάχημαι.
οὐ μέν σφεας ἔτ' ἔολπα μίνυνθά γε κειέμεν οὕτως
καὶ μάλα περ φιλέοντε· τάχ' οὐκ ἐθελήσετον ἄμφω
εὕδειν· ἀλλά σφωε δόλος καὶ δεσμὸς ἐρύξει,
εἰς ὅ κέ μοι μάλα πάντα πατὴρ ἀποδῷσιν ἔεδνα,
ὅσσα οἱ ἐγγυάλιξα κυνώπιδος εἵνεκα κούρης,
οὕνεκά οἱ καλὴ θυγάτηρ, ἀτὰρ οὐκ ἐχέθυμος."
ἦλθε Ποσειδάων γαιήοχος, ἦλθ' ἐριούνης
Ἑρμείας, ἦλθεν δὲ ἄναξ ἑκάεργος Ἀπόλλων.
θηλύτεραι δὲ θεαὶ μένον αἰδοῖ οἴκοι ἑκάστη.
ἔσταν δ' ἐν προθύροισι θεοί, δωτῆρες ἑάων·
ἄσβεστος δ' ἄρ' ἐνῶρτο γέλως μακάρεσσι θεοῖσι
τέχνας εἰσορόωσι πολύφρονος Ἡφαίστοιο.
ὧδε δέ τις εἴπεσκεν ιδὼν ἐς πλησίον ἄλλον·
"οὐκ ἀρετᾷ κακὰ ἔργα· κιχάνει τοι βραδὺς ὠκύν,

ὡς καὶ νῦν Ἥφαιστος ἐὼν βραδὺς εἷλεν Ἄρηα
ὠκύτατόν περ ἐόντα θεῶν οἳ Ὄλυμπον ἔχουσιν,
χωλὸς ἐὼν τέχνῃσι· τὸ καὶ μοιχάγρι' ὀφέλλει."
ὣς οἱ μὲν τοιαῦτα πρὸς ἀλλήλους ἀγόρευον·
'Ερμῆν δὲ προσέειπεν ἄναξ Διὸς υἱὸς 'Απόλλων·
"Ερμεία, Διὸς υἱέ, διάκτορε, δῶτορ ἑάων,
ἦ ῥά κεν ἐν δεσμοῖς ἐθέλοις κρατεροῖσι πιεσθεὶς
εὕδειν ἐν λέκτροισι παρὰ χρυσέῃ 'Αφροδίτῃ,"
τὸν δ' ἠμείβετ' ἔπειτα διάκτορος ἀργεϊφόντης·
"αἲ γὰρ τοῦτο γένοιτο, ἄναξ ἑκατηβόλ' 'Απολλον·
δεσμοὶ μὲν τρὶς τόσσοι ἀπείρονες ἀμφὶς ἔχοιεν,
ὑμεῖς δ' εἰσορόῳτε θεοὶ πᾶσαί τε θέαιναι,
αὐτὰρ ἐγὼν εὕδοιμι παρὰ χρυσέῃ 'Αφροδίτῃ."
ὣς ἔφατ', ἐν δὲ γέλως ὦρτ' ἀθανάτοισι θεοῖσιν.
οὐδὲ Ποσειδάωνα γέλως ἔχε, λίσσετο δ' αἰεὶ
Ἥφαιστον κλυτοεργὸν ὅπως λύσειεν Ἄρηα.
καί μιν φωνήσας ἔπεα πτερόεντα προσηύδα·
"λῦσον· ἐγὼ δέ τοι αὐτὸν ὑπίσχομαι, ὡς σὺ κελεύεις,
τίσειν αἴσιμα πάντα μετ' ἀθανάτοισι θεοῖσιν."
τὸν δ' αὖτε προσέειπε περικλυτὸς ἀμφιγυήεις·
"μή με, Ποσείδαον γαιήοχε, ταῦτα κέλευε·
δειλαί τοι δειλῶν γε καὶ ἐγγύαι ἐγγυάασθαι.
πῶς ἂν ἐγώ σε δέοιμι μετ' ἀθανάτοισι θεοῖσιν,
εἴ κεν Ἄρης οἴχοιτο χρέος καὶ δεσμὸν ἀλύξας;"
τὸν δ' αὖτε προσέειπε Ποσειδάων ἐνοσίχθων·
"Ἥφαιστ', εἴ περ γάρ κεν Ἄρης χρεῖος ὑπαλύξας
οἴχηται φεύγων, αὐτός τοι ἐγὼ τάδε τίσω."
τὸν δ' ἠμείβετ' ἔπειτα περικλυτὸς ἀμφιγυήεις·
"οὐκ ἔστ' οὐδὲ ἔοικε τεὸν ἔπος ἀρνήσασθαι."
ὣς εἰπὼν δεσμὸν ἀνίει μένος 'Ηφαίστοιο.
τὼ δ' ἐπεὶ ἐκ δεσμοῖο λύθεν, κρατεροῦ περ ἐόντος,
αὐτίκ' ἀναΐξαντε ὁ μὲν Θρῄκηνδε βεβήκει,
ἡ δ' ἄρα Κύπρον ἵκανε φιλομμειδὴς 'Αφροδίτη,
ἐς Πάφον· ἔνθα δέ οἱ τέμενος βωμός τε θυήεις.
ἔνθα δέ μιν Χάριτες λοῦσαν καὶ χρῖσαν ἐλαίῳ
ἀμβρότῳ, οἷα θεοὺς ἐπενήνοθεν αἰὲν ἐόντας,
ἀμφὶ δὲ εἵματα ἕσσαν ἐπήρατα, θαῦμα ἰδέσθαι.

Meanwhile the bard began to sing[3] the loves of Ares and Aphrodite, and how they first began their intrigue in the house of Hephaestus. Ares made Aphrodite many presents, and defiled King Hephaestus's marriage bed, so the sun, who saw what they were about, told Hephaestus. Hephaestus was

[3] Verse 266. αείδειν (= *to sing*): In verses 266–369, Demodocus, the bard, sings of the affair of Ares and Aphrodite and how Hephaestus had them bound.

very angry when he heard such dreadful news, so he went to his smithy brooding mischief, got his great anvil into its place, and began to forge some chains which none could either unloose or break, so that they might stay there in that place. When he had finished his snare he went into his bedroom and festooned the bed-posts all over with chains like cobwebs; he also let many hang down from the great beam of the ceiling.[4] Not even a god could see them, so fine and subtle were they. As soon as he had spread the chains all over the bed, he made as though he were setting out for the fair state of Lemnos, which of all places in the world was the one he was most fond of. But Ares kept no blind look out, and as soon as he saw him start, hurried off to his house, burning with love for Aphrodite.

Now Aphrodite was just come in from a visit to her father Zeus,[5] and was about sitting down when Ares came inside the house, an said as he took her hand in his own, "Let us go to the couch of Hephaestus: he is not at home, but is gone off to Lemnos among the Sintians, whose speech is barbarous."

She was nothing loth, so they went to the couch to take their rest, whereon they were caught in the toils which cunning Hephaestus had spread for them, and could neither get up nor stir hand or foot, but found too late that they were in a trap. Then Hephaestus came up to them, for he had turned back before reaching Lemnos, when his scout the sun told him what was going on. He was in a furious passion, and stood in the vestibule making a dreadful noise as he shouted to all the gods.

"Father Zeus," he cried, "and all you other blessed gods who live for ever, come here and see the ridiculous and disgraceful sight that I will show you. Zeus's daughter Aphrodite is always dishonoring me because I am lame. She is in love with Ares, who is handsome and clean built, whereas I am a cripple-but my parents are to blame for that, not I; they ought never to have begotten me. Come and see the pair together asleep on my bed. It makes me furious to look at them. They are very fond of one another, but I do not think they will lie there longer than they can help, nor do I think that they will sleep much; there, however, they shall stay till her father has repaid me the sum I gave him for his baggage of a daughter, who is fair but not honest."

On this the gods gathered to the house of Hephaestus. Earth-encircling Poseidon came, and Hermes the bringer of luck, and King Apollo, but the goddesses stayed at home all of them for shame. Then the givers of all good things stood in the doorway, and the blessed gods roared with inextinguishable laughter, as they saw how cunning Hephaestus had been, whereon one would

---

[4] Verse 279, μελαθρόφιν (*from the roof's beam*): μέλαθρον (*mélathron*) is the large beam supporting the roof. It came to mean the roof and finally the whole of the house. Since it was smoked (the fire-place was lying underneath, therefore it was blackened (= μέλαινα)). Eventually, μέλαθρον ended up meaning *mansion* or *palace*.

[5] Verse 289. πατρός (= *father's*): According to Homer, Aphrodite was the daughter of Zeus Cronides (Cronus' son) and Dione. Another myth wants her to have emerged out of the froth of sea waves in Cyprus or in the island of Cethyra, therefore, she is named Κυπρογενής (*Cyprus-born*) or Κυθερείη (*Cythereia* = *of Cethyra*).

Figure 12.3 Gaetano Gadolfi (1734–1802): Aphrodite in Hephaestus' Forge.

turn towards his neighbor saying: "Ill deeds do not prosper, and the weak confound the strong. See how limping Hephaestus, lame as he is, has caught Ares who is the fleetest god in heaven; and now Ares will be cast in heavy damages."

Thus did they converse, but King Apollo said to Hermes, "Messenger Hermes, giver of good things, you would not care how strong the chains were, would you, if you could sleep with Aphrodite?"

"King Apollo," answered Hermes, "I only wish I might get the chance, though there were three times as many chains – and you might look on, all of you, gods and goddesses, but would sleep with her if I could."

The immortal gods burst out laughing as they heard him, but Poseidon took it all seriously, and kept on imploring Hephaestus to set Ares free again. "Let him go," he cried, "and I will undertake, as you require, that he shall pay you all the damages that are held reasonable among the immortal gods."

> "Do not," replied Hephaestus, "ask me to do this; a bad man's bond is bad security; what remedy could I enforce against you if Ares should go away and leave his debts behind him along with his chains?"
>
> "Hephaestus," said Poseidon, "if Ares goes away without paying his damages, I will pay you myself." So Hephaestus answered, "In this case I cannot and must not refuse you."
>
> Thereon he loosed the bonds that bound them, and as soon as they were free they scampered off, Ares to Thrace and laughter-loving Aphrodite to Cyprus and to Paphos, where is her grove and her altar fragrant with burnt offerings. Here the Graces hathed her, and anointed her with oil of ambrosia such as the immortal gods make use of, and they clothed her in raiment of the most enchanting beauty.
>
> (Od. 8.266–366)

The following remarks pertain to the two traps: (a) There was an activation mechanism of the trap, triggered by the weight of the person sitting or lying on the respective piece of furniture: the seat was not activated by the weight of one single person, but by the weight of two persons, so that Aphrodite would not be captured when going to bed alone. (b) There was a mechanism in the bed locking the "invisible net" hanging from the roof (μελαθρόφιν), which dropped upon activation of the previous mechanism, and bound the couple so tight that they could hardly move. (c) Much more interesting that the mechanisms, is the net itself: to be "invisible" it must have been made of extremely thin and strong fibres. Such materials are rather modern technological achievements, e.g., glass and carbon fibres, or even organic fibres such as Kevlar. If such materials were available in Homer's era, undoubtedly that civilization was marked by this highly developed technology.

Further Constructions by Hephaestos ('Ηφαιστότευκτα = Hephaestus-made)

This is a more or less exhaustive list of Hephaestus' various constructions, some of which are analyzed in the following chapters: The Olympian mansions, Zeus' scepter, Zeus'and Athena's aegis, Athena's spear, Apollo's chariot and Demeter's sickle, the armours of the gods in their war against the Titans, Helios' golden bed, Apollo's and Artemis' bows and arrows, Agamemnon's sceptre, Aphrodite's golden girdle, Zeus' hound, Alcinous' mastiffs, necklace of Harmonia, Aeëtes' bulls, Aeëtes' fountains, Talos of Crete, Hephaestus' golden handmaidens, Ariadne's wreath, Golden vineyard, Dionysus' craters, Perseus' scythe, Heracles' rattles, Heracles' golden breastplate, Achilles' weapons including the famous shield, Helios' crater.

Figure 12.4 Giordano Luca (1634–1705): Aphrodite, Ares and the forge of Hephaestus, Palazzo Ruspoli, Rome (reproduced by permission).

Figure 12.5 Giordano Luca (1634–1705): Ares and Aphrodite captured by Hephaestus, 1670, Lichtenstein Museum, Vienna (reproduced by permission).

Chapter 13
The Robots of Hephaestus

Hephaestus, besides his automatic/autonomous devices, was capable of developing much more advanced and complex machines, in fact, even the most evolved device of modern technology: Robots.

A robot is a self-governed, programmable electromechanical apparatus, applied both in industry and in scientific research, performing specific tasks or a limited variety of tasks. The robots belong to the class of automated devices. There are no generally accepted definitions of robots as compared with other automatic apparatuses. Simply, they are characterized by an improved adaptaptability and re-programmability, rendering them suitable to perform jobs, mainly repetitive, faster, less expensively and with improved precision as compared with humans. Moreover, they can operate in adverse or hazardous environments, either industrial or in great sea depths or, finally, in space.

The term "robot" originates from the Czech word *robota*, meaning *forced labour*. It was coined through *R.U.R.*,[1] human society fully depending on mechanical workers, called "robots", who had the ability to perform all kinds of mental work or labour (Figure 13.1). It is intresting that, through this play, the word "robot" was introduced to the *Oxford English Dictionary*, and through it to the rest of the world languages. Nowadays, the word, along with its companion word "Robotics", meaning a branch of technology dealing with the construction and operation of robots, have become very common, since, due to the rapidly advancing technology, tend to be part of every day life.

[1] *Rossum's Universal Robots*, Oxford University Press, London, 1927, Greek translation by S.A. Paipetis (to be published).

DOI 10.1007/978-90-481-2514-2_13, © Springer Science + Business Media B.V. 2010

Figure 13.1 Costumes of R.U.R. at its first performance in 1922.

Robots are mentioned in the Iliad for the first time in human history: This is a remark by Isaac Asimov,[2] who, within a total of 17 short stories, published mainly between 1940 and 1960, analyzed the problems and social effects caused by the wide use of robots and stated the famous "three laws of robotics", i.e., the conditions under which robots can be used efficiently and safely by humans.[3]

Science fiction and popular fantasy presents robots as "mechanical men" or "androids" (from the Greek word ἄνδρας, i.e., *man*). This is not always the case, since the form of a robot corresponds to the particular kind of work it has to perform. So, there is a basic class of automated machinery, called *robota*, executing specific repetitive jobs, for example, on a car production line (Figure 13.2a). Futher examples appear in Figures 13.2b and c.

On the contrary, the greater diversity of jobs is required from a robot, the more its form tends to become human (Figures 13.2b and c), for a simple reason: The human body, through its age-long adaptive evolution within the terrestrial environments, is the perfect "universal tool", therefore, if robots are to substitute for humans in their activities, they must assume their form.

(a) *The gold-and-silver mastiffs of Alcinous*

In the Odyssey, two guardian-mastiffs made of gold and silver, made by Hephaestus, were the doorkeepers of the magnificent palace of king Alcinous in Phaeacia. The task assigned to these robots was neither too simple

[2] Isaac Asimov (1920–1992), a famous science fiction writer and also popularized science books in nearly all fields of science.

[3] Asimov, Isaac, *I, Robot* and *The Rest of the Robots*, Panther Science Fiction, 1977.

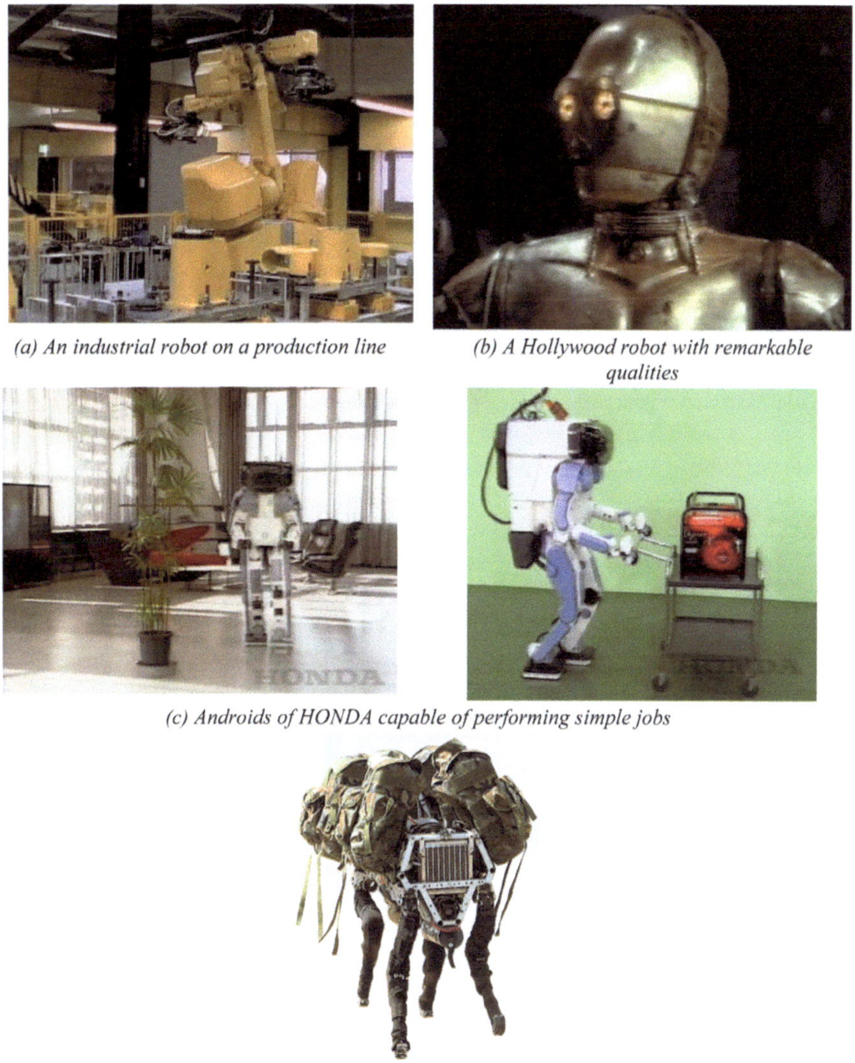

(a) An industrial robot on a production line

(b) A Hollywood robot with remarkable qualities

(c) Androids of HONDA capable of performing simple jobs

(d) The BigDog, a military robot prototype developed by Boston Dynamics with admirable abilities.

Figure 13.2 Various types of robots for different applications.

nor repetitive, but fairly complicated. In fact, they had both to simulate real mastiffs in all of their reactions as guardians of the palace, and remain invulnerable against assaults that real animals would not be able to face. The account is as follows:

> Ὣς ἄρα φωνήσασ' ἀπέβη γλαυκῶπις Ἀθήνη
> πόντον ἐπ' ἀτρύγετον, λίπε δὲ Σχερίην ἐρατεινήν,

ἵκετο δ' ἐς Μαραθῶνα καὶ εὐρυάγυιαν 'Αθήνην,
δῦνε δ' 'Ερεχθῆος πύκινον δόμον, αὐτὰρ 'Οδυσσεὺς
'Αλκινόου πρὸς δώματ' ἴε κλυτά· πολλὰ δὲ οἱ κῆρ
ὅρμαιν' ἱσταμένῳ, πρὶν χάλκεον οὐδὸν ἱκέσθαι.
ὥς τε γὰρ ἠελίου αἴγλη πέλεν ἠὲ σελήνης
δῶμα κάθ' ὑψερεφὲς μεγαλήτορος 'Αλκινόοιο.
χάλκεοι υὲν γὰρ τοῖχοι ἐληλέδατ' ἔνθα καὶ ἔνθα,
ἐς μυχὸν ἐξ οὐδοῦ, περί δε θριγκὸς κυάνοιο·
χρύσειαι δὲ θύραι πυκινὸν δόμον ἐντὸς ἔεργον·
σταθμοὶ δ' ἀργύρειο ἐν χαλκέῳ ἔστασαν οὐδῶ,
ἀργύρεον δ' ἐφ' ὑπερθύριον, χρυσέη δὲ κορώνη.
χρύσειοι δ' ἑκάτερθε καὶ ἀργύρεοι κύνες ἦσαν,
οὕς Ἥφαιστος ἔτευξεν ἰδυίῃσι πραπίδεσσι
δῶμα φυλασσέμεναι μαγαλήτορος 'Αλκινόοιο,
ἀθανάτους ὄντας καὶ ἀγήρως ἤματα πάντα.

Then Athena left Scheria and went away over the sea. She went to Marathon and to the spacious streets of Athens, where she entered the abode of Erechtheus; but Ulysses went on to the house of Alcinous, and he pondered much as he paused a while before reaching the threshold of bronze, for the splendour of the palace was like that of the sun or moon. The walls on either side were of bronze from end to end, and the cornice was of blue enamel. The doors were gold, and hung on pillars of silver that rose from a floor of bronze, while the lintel was silver and the hook of the door was of gold. **On either side there stood gold and silver mastiffs which Vulcan, with his consummate skill, had fashioned expressly to keep watch over the palace of king Alcinous; so they were immortal and could never grow old.**

(Od. 7.78–94)

(b) *The golden girls of Hephaestus*

Two robots in the form of "female" robots were the personal assistants of Hephaestus, who supported him with his work in the forge and "had voice and sense":

ἦ, καὶ ἀπ' ἀκμοθέτοιο πέλωρ αἴητον ἀνέστη
χωλεύων· ὑπὸ δὲ κνῆμαι ῥώοντο ἀραιαί,
φύσας μὲν ῥ' ἀπάνευθε τίθει πυρός, ὅπλά τε πάντα
λάρνακ' ἐς ἀργυρέην συλλέξατο, τοῖς ἐπονεῖτο;
σπόγγῳ δ' ἀμφὶ πρόσωπα καὶ ἄμφω χεὶρ ἀπομόργνυ
αὐχένα τε στιβαρὸν καὶ στήθεα λαχνήεντα,
δὺ δὲ χιτών', ἕλε δὲ σκῆπτρον παχύ, βῆ δὲ θύραζε
χωλεύων· ὑπὸ δ' ἀμφίπολοι ρώοντο ἄνακτι
χρύσειαι ζωῄσι νεήνισιν εἰοικυῖαι.
τῆς ἐν μὲν νοὸς ἐστὶ μετὰ φρεσίν, ἐν δὲ καὶ αὐδὴ
καὶ σθένος, ἀθανάτων δὲ θεῶν ἄπο ἔργα ἴσασιν.

αἳ μὲν ὕπαιθα ἄνακτος ἐποίπνυον· αὐτὰρ ὁ ἔρρων
πλησίον, ἔνθα Θέτις περ, ἐπὶ θρόνου ἷζε φαεινοῦ,
ἐν τ' ἄρα οἵ φῦ χειρὶ ἔπος τ' ἔφατ' ἔκ τ' ὀνόμαζε.

> On this the mighty monster hobbled off from his anvil, his thin legs plying lustily under him. He set the bellows away from the fire, and gathered his tools into a silver chest. Then he took a sponge and washed his face and hands, his shaggy chest and brawny neck; he donned his shirt, grasped his strong staff, and limped towards the door. There were golden handmaids also who worked for him, and were like real young women, with sense and reason, voice also and strength, and all the learning of the immortals; these busied themselves as the king bade them, while he drew near to Thetis, seated her upon a goodly seat, and took her hand in his own, saying, . . .
>
> (Il. 18.410–423)

Homer provides no further details of the miraculous devices, neither can be unreservedly assumed that robots were available, at a time that neither prime movers, nor electricity and, of course, no computers or artificial intelligence are known to exist. However, even though a poetic conception, this account expresses admirably an inherent desire of humans from time immemorial, as well as the boundless abilities of human mind and hands.

Chapter 14
The Ships of the Phaeacians and the UAVs

Odysseus, shipwrecked in Phaeacia, heard of the ships of the Phaeacians for the first time from goddess Athena, who, disguised as peasant girl, informed him of local people and their habits. So, the Phaeacians:

> νηυσὶ θοῇσιν τοί γε πεποιθότες ὠκείῃσι
> λαῖτμα μέγ' ἐκπερόωσιν, ἐπεί σφισι δῶκ' ἐνοσίχθων·
> τῶν νέες ὠκεῖαι ὡς εἰ πτερὸν ἠέ νόημα.
>
> They are a sea-faring folk, and sail the seas by the grace of Neptune in ships that glide along like thought, or as a bird in the air.
>
> (Od. 7.34–36)

On another occasion, the ship offered by king Alcinous to Odysseus, to take him from Phaeacia[1] to Ithaca, and of its propulsion is presented through an amazing account:

[1] According to Schliemann, Phaeacia was the island of Kerkyra (or Corfu), a view seriously questioned by many interpretors of the various accounts of Odyssey. An explicit reference in the *Argonautics* of the Orphic texts (1291–1297) though, referring to times prior to the Trojan War, but probably later than the Homeric Epics, is as follows:

> Αὔταρ ἐπεὶ καὶ τόνδε πότμον παράμειψε θέουσα
> 'Αργῶ, κύμα δὲ πόντου καὶ κόλπον ἵκανε,
> λαιψηροῖς πλήθουσα κατὰ προτόνων ἀνέμοισι,
> Κέρκυραν ζεθέην ἐξίκετο, τὴν σφὶν ἔναιον
> ἴδριες εἰρεσίης καὶ ἀλιπλάγκτοιο πορείης
> Φαίηκες· τοῖσιν δ' ἀρ' ἐφημοσύναισι θέμιστας
> 'Αλκίνοος κραίνεσκε δικαιότατος βασιλήων.
>
> Argo (the ship of the Argonauts), having escaped this disaster too by rushing on the sea waves, entered a bay, with sails fully bloated by the strong winds and arrived at the divine Kerkyra, inhabited by the Phaeacians,

S.A. Paipetis, *The Unknown Technology in Homer*, History of Mechanism and Machine Science 9,
DOI 10.1007/978-90-481-2514-2_14, © Springer Science + Business Media B.V. 2010

... τοι δὲ καθῖζον ἐπὶ κληῖσιν ἕκαστοι
κόσμῳ, πεῖσμα δ' ἔλυσαν ἀπὸ τρητοῖο λίθοιο.
εὖθ' οἱ ἀνακλινθέντες ἀνερρίπτουν ἅλα πηδῷ,
καὶ τῷ νήδυμος ὕπνος ἐπὶ βλεφάροισιν ἔπιπτε,
νήγρετος ἥδιστος, θανάτῳ ἄγχιστα ἐοικώς.
ἡ δ', ὥς τ' ἐν πεδίῳ τετράοποι ἄρσενες ἵπποι,
πάντες ἅμ' ὁρμηθέντες ὑπὸ πληγῇσιν ἱμάσθλης
ὑψόσ' ἀειρόμεναι ῥίμφα πρήσσουσιν κέλευθον,
ὣς ἄρα τῆς πρύμνη μὲν ἀείρετο, κῦμα δ' ὄπισθε
πορφύρεον μέγα θῦε πολυφλοίσβοιο θαλάσσης.
ἡ δὲ μάλ' ἀσφαλέως θέεν ἔμπεδον· οὐδὲ κὲν ἴρηξ
κίρκος ὁμαρτήσειεν, ἐλαφρότατος πετεηνῶν·
ὣς ἡ ῥίμφα θέουσα θαλάσσης κύματ' ἔταμνεν,
ἄνδρα φέρουσα θεοῖσ' ἐναλίγκια μήδε' ἔχοντα,
ὅς πρὶν μὲν μάλα πολλὰ πάθ' ἄλγεα ὅν κατὰ θυμόν,
ἀνδρῶν τε πτολέμους ἀλεγεινά τε κύματα πείρων·
δὴ τότε γ' ἀτρέμας εὗδε, λελασμένος ὅσσ' ἐπεπόνθει.

Then he too went on board and lay down without a word, but the crew took every man his place and loosed the hawser from the pierced stone to which it had been bound. Thereon, when they began rowing out to sea, Ulysses fell into a deep, sweet, and almost deathlike slumber. The ship bounded forward on her way as a four in hand chariot flies over the course when the horses feel the whip. Her prow curveted as it were the neck of a stallion, and a great wave of dark blue water seethed in her wake. She held steadily on her course, and even a falcon, swiftest of all birds, could not have kept pace with her. Thus, then, she cut her way through the water. carrying one who was as cunning as the gods, but who was now sleeping peacefully, forgetful of all that he had suffered both on the field of battle and by the waves of the weary sea.

(Od. 13.76–92)

Concerning the ship, the description corresponds to the so-called "flying boats", which are lifted above the sea surface, equipped with four floaters (hydrofoils) remaining in the water. This minimizes flow resistance and leads to high speeds. Hydrofoils can use either propellers or jet propulsion (Figure 14.1).

---

unrivalled with oars and rough seas, over who Alcinous, the most righteous king, issues the laws by his resolutions.

Besides, Apollodorus in his *Argonautics*, referring to the Argonauts, writes: Πραπεμψάμενοι δὲ θρινακίαν νῆσον 'Ηλίου βόας ἔχουσαν είς ιὴν Παιάκων νῆσον Κέρκυραν ἧκον, ἧς βασιλεὺς ἦν 'Αλκίνοος (= *Having passed by Thrinacia island, where the cattle of Helios were, they arrived at the Island of the Phaeacians, Kerkyra, whose king was Alcinous*).

Figure 14.1 A US Navy hydrofoil.

Figure 14.2 American Navy Combat screw-propelled hovercraft.

The same description, although for less speed, applies to *hovercrafts*, i.e., land or sea vehicles moving on an air cushion, produced by huge blowers, pushing air downwards and raise the whole structure above land or sea surface, where the vehicle can move with equal ease. They are using propellers and are applied both as transport or military vehicles, for landing or for combat (Figure 14.2).

Finally, king Alcinous addresses Odysseus as follows:

εἰπὲ μοι γαῖάν τε τεὴν δῆμον τε πόλιν τε,
ὄφρα σε τῇ πέμψωσι τιτυσκόμεναι φρεσὶ νῆες.
οὐ γὰρ Φαίηκεσσι κυβερνητῆες ἔασιν,
οὐδὲ τι πηδάλι' ἐστί, τά τ' ἄλλαι νῆες ἔχουσιν·
ἀλλ' αὐταὶ ἴσασι νοήματα καὶ φρένας ἀνδρῶν,
καὶ πάντων ἴσασι πόλιας καὶ πίονας ἀγροὺς
ἀνθρώπων, καὶ λαῖτμα τάχισθ' ἁλὸς ἐκπερόωσιν
ἠέπι καὶ νεφέλῃ κεκαλυμμέναι· οὐδὲ ποτὲ σφιν
οὔτε τι πημανθῆναι ἔπι δέος οὔτ' ἀπολέσθαι.
ἀλλὰ τόδ' ὥς πότε πατρὸς ἐγὼν εἰπόντος ἄκουσα
Ναυσιθόου, ὅς ἔφασκε Ποσειδάων' ἀγάσεσθαι
ἡμῖν, οὕνεκα πομποὶ ἀπήμονές εἰμεν ἁπάντων·
φῆ ποτὲ Φαιήκων ἀνδρῶν εὐεργέα νῆα
ἐκ πομπῆς ἀνιοῦσαν ἐν ἠεροειδέϊ πόντῳ
ῥαισέμεναι, μέγα δ' ἧμιν ὄρος πόλει ἀμφικαλύψειν,
ὥς ἀγόρευ' ὁ γέρων· τὰ δέ κεν θεὸς ἤ τελέσειεν,
ἤ κ' ἀτέλεστα εἴη, ὥς οἱ φίλον ἔπλετο θυμῷ.

Tell me also your country, nation, and city, that our ships may shape their purpose accordingly and take you there. For the Phaeacians have no pilots; their vessels have no rudders as those of other nations have, but the ships themselves understand what it is that we are thinking about and want; they know all the cities and countries in the whole world, and can traverse the sea just as well even when it is covered with mist and cloud, so that there is no danger of being wrecked or coming to any harm. Still I do remember hearing my father Nausithous say that Neptune was angry with us for being too easy-going in the matter of giving people escorts. He said that one of these days he should wreck a ship of ours as it was returning from having escorted some one, and bury our city under a high mountain. This is what my father used to say, but whether the god will carry out his threat or no is a matter which he will decide for himself.

(Od. 8.555–571)

Therefore, the ships of the Phaeacians possess no steering system or other device to control motion and directivity, and no crew was necessary. It is to be noted, that the 'Απήδαλος Ναῦς, e.g., *a ship without rudder*, is the present-time crest of the island of Kerkyra (or Corfu), the Homeric Island of the Phaeacians, inspired by this description.

Moreover, the ships are intelligent and, once they know their destination, there they go themselves. This may be a beautiful fairy tale, however, in modern language, it can be stated that the ships were centrally controlled by an artificial intelligence system.

Modern air vehicles so described are termed as "Unmanned Air Vehicles" (UAVs). They are not remote-controlled, but they possess artificial intelli-

Figure 14.3 The rudderless ship, depicted on the keystone of the Town Hall of the City of Corfu.

gence, being capable, while searching for predetermined targets, to make decisions for the best possible completion of the task.

Such air-vehicles may be of various kinds, size and design, depending on the particular task, mainly for military opercations: there may be from being bombers to reconnaissance or spy aircrafts or even for street-fighting operations. They may be as small as 25 cm and can accomplish a wide range of missions. For example, the instructions might be to attach itself to an external wall, "listen" to what is said inside and transmit the information. Or to enter interior spaces and measure contaminants in the atmosphere in ppm. Or wait patiently before a locked door, and jump in when somebody opens from the inside.

These aircrafts are the latest development of military technology. They are made of extremely light and strong materials and operate in extreme environments of external pressure, temperature and humidity, even corrosive or toxic. They carry exotic propellants, providing them with energy, and are equipped with turbo-propellers, as well as generators to support their control and telecommunications systems, as well as computers with huge memories, carriers of their artificial intelligence. Finally, they carry a payload of explosives or laboratory equipment, for specific purposes (Figure 14.4).

The Homeric account applies to an unmanned passenger sea-vehicle, probably a high speed jet hydrofoil equipped with all devices of UAVs. Could such advanced ecquipment be available at the end of Bronze Age or at the

Eagle Eye (Bell Helicopter Textron Incorporation)

RQ-1 Predator MAE UAV (General Atomics)

The Outrider Tactical UAV (Alliant Techsystems)

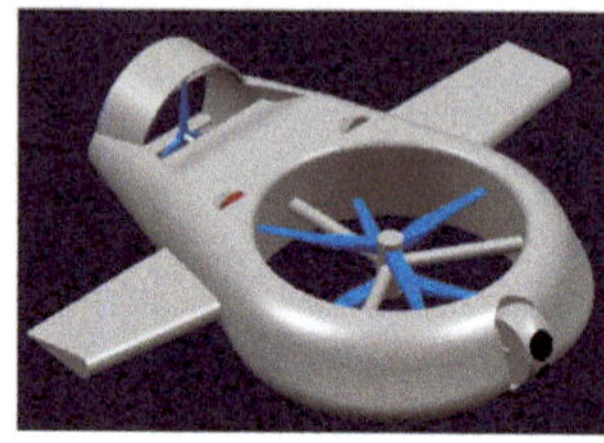

Dragon Warrior (vertical take-off)

Figure 14.4 Modern UAV fighters.

dawn of the Iron Age? Unfortunately, no further evidence is available, as is the case with other structures, to be examined later. However, such miraculous conceptions, not even mentioned as the work of some god, show that this kind of ideas pre-existed in the thoughts of Mycenaean Greeks, for almost three millenia before their appearance in the modern world.

Part 4
Defensive Weapons in the Epics

Chapter 15
Structural Materials and Analytical Processes

This is an introduction to the next three chapters, in which very interesting structures are dealt with, exhibiting impressive similarity with advanced modern structures. These are the shields of Achilles and Ajax, consisting of successive laminates, metallic for the former, of calf's leather for the latter. Such elements are called *multi-layered structures*, belonging to the wider class of *composite materials*. The present chapter deals both with materials, in Homer's era and nowadays, and the numerical methods applied for the analysis of the said structures and their operation.

15.1 Metals in Homer

From the point of view of technology, early civilizations are classified according to the metal available at the time, such as Copper/Bronze or Iron Age. Probably, a real "Copper Age" never existed, except perhaps a short period at the beginning of the Egyptian Civilization. Copper was a soft metal with limited applicability, mainly used for coins and jewellery. To manufacture tools and weapons, hardening techniques were needed. This is how alloys were invented. Copper was melted with other metals, such as lead, antimony and arsenic; bronze in its numerous forms was produced by the reduction of certain mineral mixtures. The most popular and well-known bronze was made from copper and tin at 10:1 proportion. It was a yellowish metal that could be melted and cast to the desired form. Copper and gold workers passed along to bronze workers the technique of melting the material at elevated temperature and casting it in simple clay or stone moulds, to produce axe or spear heads or other solid objects. The production technique of hollow receptacles or sculptures consisted of the preparation of a wax

S.A. Paipetis, *The Unknown Technology in Homer*, History of Mechanism and Machine Science 9, DOI 10.1007/978-90-481-2514-2_15, © Springer Science + Business Media B.V. 2010

model subsequently encased in a clay mould. At high temperature, the wax melted and flowed out, leaving a cavity behind, in which the molten metal was cast.

In this way, bronze became the most important material for early civilizations and its uninhibited supply was the object of countless arrangements and treaties between coutries. In the aluvial valleys of rivers, where various civilizations were developed, metals were rare, and had to be transported from remote mining areas. Tin was very rare in the Mid-East, and Bronze Age civilizations had to search for it far beyond their boundaries. It is worth noting that, knowledge of Eastern civilizations was gradually moving to the West, along the rapidly developing trade routes.

The Homeric Epics are believed to belong to the Bronze era. The metal is mentioned as copper 128 times in the Iliad against 23 mentions of iron (mostly describing qualitative characteristics, such as *iron heart* or *iron courage*), while in the Odyssey, shorter than the Iliad by 3,500 lines, is mentioned 28 times only against 22 mentions of iron (5 qualitative).

Detailed references to metals in Homer are given by Zeggelis[1] and subsequent writers.[2] In particular, for the present, metals of interest are bronze, tin, pure gold (Achilles' shield) and also bronze and calf's leather (Ajax's shield). The physical properties of these materials are presented in Table 15.1.

15.2 Composite Materials

In modern technology, a very important role is played by *composite materials*, i.e., materials generated by two or more simple or *monolithic* materials that exhibit substantially improved properties, not exhibited by any of the constituent materials alone. The latter are *mechanically* combined, i.e., as if they were bonded together. They do not interact chemically, i.e., are not classified as chemical compounds and also they do not dissolve within one another, i.e., they are neither solutions nor metallic alloys.

---

[1] K.D Zeggelis, *The Science of Nature in Homer*, Athens 1891, republished in 1987 by University of Patras Editions, introductory essay by S.A. Paipetis (in Greek).

[2] G. Varoufakis, Iron in the Homeric Epics, in *Science and Technology in the Homeric Epics*, S.A. Paipetis (Ed.), Springer, 2008.

Table 15.1 Properties of materials,[3] used for the numerical and experimental simulation of Achilles' and Ajax's shields.

Material and Properties	Unit	SAE 40 Bronze (shields)	Tin (Achilles' shield)	Gold (Achilles' shield)	Phosphor Bronze, annealed (projectile)	Calf's leather (Ajax's shield)
Tensile strength	MPa	255–300	27.0	10.79	276	–
E-modulus	GPa	93	42	78	110–117	1.0
Elongation at break	%	3.00–4.38	40.0	50.00	3.80%	–
Vicker's hardness	MPa	870		216		–
Poisson's ratio	–	0.33	0.36	0.44	0.31	0.29
Density	kg/dm^3	8.90	7.29	19.29	8.90	–

Composite materials usually consist of a basic *matrix* material, in which particles, grains, fibres of glass, plastic, metal, carbon, etc., are embedded, either to enhance the physical properties of the matrix, to facilitate the production process or to reduce production cost. Depending on the specific application, the matrix may be metallic, polymeric (plastic) or ceramic. Well-known examples are fibre glass (consisting of a resin, epoxide or polyester resin matrix and glass fibres, a suitable material for recreation boats) and vulcanized rubber for numerous other applications.

Advanced composites, responding to high operational requirements, such as high strength, light weight, durability, stability at high temperature or in aggressive environments etc., are used for load-bearing parts of land, air, sea or space vehicles, advanced machinery, etc. (Figures 15.1 to 15.3). They are mainly *laminated* materials, consisting of thin successive layers with widely different properties. Other applications include sports equipment, such as tennis rackets, poles for pole jumping, motorist helmets (Figure 15.4), etc., as well as transportation equipment from travel suitcases up to big containers, with lighter weights decreasing transportation cost.

It is amazing that, for the first time in human history, the idea of laminated structures appears in the Iliad, in the shields of Achilles and Ajax, with such structural details that, based on realistic assumptions, their reconstruction and investigation are possible, both numerically and, by means of modern

[3] The *kinetic energy* of a spear and of an air-gun projectile, applied for the numerical and experimental simulation of shields' behaviour, was equal to *39.38 Joule*, corresponding to the kinetic energy of a javelin thrown by the world recordman, year 2000.

(a)

(b)

(c)

Figure 15.1 (a) Aglider, (b) an F-18 Hornet fighter and (c) a Boeing 787 passenger aircraft are all made, by a great percentage, of advanced composites.

computer codes, experimentally. Also their battle behaviour, as described by Homer, can be confirmed most accurately.

Till present it was thought that the earliest reference to composite materials was in the Bible (Exodus 5, 15–18), whereby the Israelites were complaining to Pharaoh, for not being provided with sufficient quantities of straw to carry out the job assigned to them, i.e., to produce bricks:

> Then the Israelite foremen came and made this appeal to Pharaoh: "Why do you treat your servants in this manner? No straw is supplied to your servants, and still we are told to make bricks. Look how your servants are beaten! It is you who are at fault." Pharaoh answered, "It is just because you are lazy that you keep saying, 'Let us go and offer sacrifice to the Lord.' Off to work, then!

(a) (b)

Figure 15.2 (a) Space Shuttle Atlantis, and (b) a modern wind generator.

(a)

(b)

Figure 15.3 (a) Hubble space telescope, and (b) a modern tank with composite ceramic armour.

Figure 15.4 Motorist helmets made of advanced composite materials.

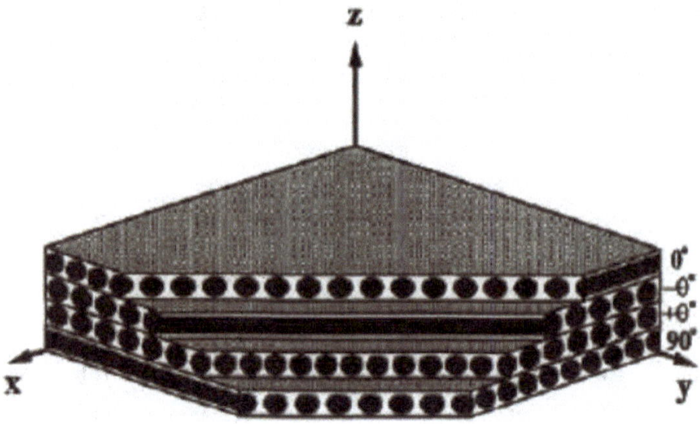

Figure 15.5 A laminated composite.

> Straw shall not be provided for you, but you must still deliver your quota of bricks."

The inner structure of a modern laminated composite (Figure 15.5) is as follows. A plane formation of parallel (unidirectional) fibres is impregnated with the matrix material, a resin. A very thin sheet, or *lamina*, is thus produced of about 1/10 mm thick. This sheet is highly *anisotropic*, i.e., its mechanical properties vary widely in various directions. To produce *laminates*, which would be *isotropic*, i.e., having practically the same properties along all directions, several sheets are bonded together at different orientations. In this way, a multi-layered laminate is produced, which, processed under suitable pressure and temperature, acquires very high thermal and mechanical strength, as well as other extraordinary properties depending on the individual application. The geometry of such a laminate appears in Figure 15.6.

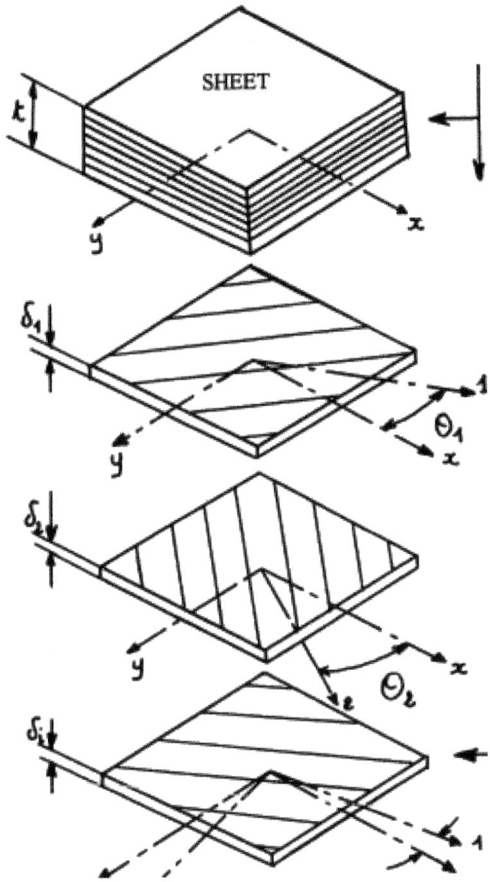

Figure 15.6 Geometry of a modern laminated composite.

Clearly, laminated structures can be produced from isotropic laminae with widely differing properties, and other modern applications. For example, two metal laminates bonded together by an intermediate plastic core, can exhibit high vibration damping, for example, in a vehicle, not only to protect its structural integrity against fatigue, but also to make the passengers feel comfortable. Such composite structures are called *sandwich structures.*

15.3 Numerical Analysis of the Contact-Impact Problem

The problem of contact-impact between solids was been particularly studied during the last three decades. Such problems are of great technological importance, being related, among others, with the simulation of high velocity impact, penetration and formation of metals, nuclear reactor safety etc. Accordingly, new algorithms, suitable for the study of contact-impact and for the theoretical formulation of nonlinear continuum mechanics problems were developed. The Finite Element method was used and, in the present, algorithms developed by the *Methods Development Group* of the distinguished center *Lawrence Livermore National Laboratory*[4] *(LLNL, University of California, USA),* properly modified. In the present analysis, the problem was formulated with special emphasis on theoretical and computational aspects of impact problems. Furthermore, the results were evaluated and conclusions drawn. Although accurate data related to the exact dimensions and mechanics of the shield are not available, information can be derived from the modern sport of javelin throwing. In fact, the kinetic energy of an arrow or spear hitting Achilles' or Ajax's shield was taken equal to that of a javelin thrown by the 2000 world recordman.

The analysis of impact behaviour of both shields is given in the sequence. The mathematical formulation of contact-impact problems is based on the theory of variational inequalities.[5] It is customary to handle such inequalities in connection with finite element processes, either via optimization techniques or by transformation to equalities, satisfying the constraint conditions through penalty or Lagrange multipliers techniques. Such formulation differs from elastodynamic problems, where non-colliding bodies are involved, since the contact discontinuity boundary condition is imposed. Therefore, upon selection of the Lagrangian formulation, the solution of the problems is based on equilibrium laws of momentum and energy along with traction, displacement and contact boundary conditions:

$$\nabla \cdot \sigma + b = \rho a \tag{15.1}$$

$$\sigma \cdot \Delta v + \nabla \cdot q + r = \dot{E} \tag{15.2}$$

$$\sigma_{ij} n_j = t_i (t) \tag{15.3}$$

[4] See D.J. Steinberg and M.W. Guinan, A High Strain Rate Constitutive Model for Metals. University of California, Lawrence Livermore National Laboratory, Report, UCRL-80465, (1978).

[5] N. Kikuchi and J.T. Oden, *Contact Problems in Elasticity*, (Society for Industrial and Applied Mathematics, Philadelphia, 1984).

$$\chi_i = \bar{\chi}_i \quad \text{on interior boundary } \partial b_2 \tag{15.4}$$

$$(\sigma_{ij}^{+} - \sigma_{ij}^{-})n_i = 0 \quad \text{on interior boundary } \partial b_3 \text{ when } \chi_i = \bar{\chi}_i \tag{15.5}$$

$$\nabla \cdot \sigma + b = \rho a$$

$$\sigma \cdot \Delta v + \nabla \cdot q + r = \dot{E}$$

$$\sigma_{ij} n_j = t_i\,(t)$$

$$\chi_i = \bar{\chi}_i \quad \text{on interior boundary } \partial b_2$$

$$(\sigma_{ij}^{+} - \sigma_{ij}^{-})n_i = 0 \quad \text{on interior boundary } \partial b_3 \text{ when } \chi_i = \bar{\chi}_i$$

where σ_{ij} is the Cauchy stress, χ_i the position vector of the deformed state at time t, η_j the unit outward normal to the boundary, ν the particle velocity, ρ the current density, b the body force, and E, q and r are the internal energy, heat flux vector per current area and internal heat generation rate, respectively. The energy balance is sometimes redundant, however, it is strongly dependent on the constitutive equations chosen for the description of the materials involved, as well as for the loading conditions. In the present case, the constitutive equations describing spear and impacted shield were elastic-plastic with hardening or elastic-plastic with thermal softening. Only in the first case isothermal conditions are implied, usually leading to small plastic strains, whereas, in the second case, an energy balance is needed to account for the localized superplastic behavior of metals, such as copper or tin, which may occur at elevated temperatures. The positive definiteness condition of the Jacobean imposes fulfillment of the continuity equation in the Lagrangian formulation, therefore, it is implicitly satisfied. With low and medium velocity impact problems, Lagrangian formulation is preferable against Eulerian. The use of interactive reasoning to overcome overdistortion of the elements at high-strain-rate applications, is one of the inherent advantages of the DYNA codes (see below) and, wherever possible, it eliminates the need to use Arbitrary Langrangian Eulerian (ALE) schemes. In the sequence, the explicit integration scheme, the contact-impact algorithm as well as the constitutive equations are briefly discussed.

15.4 Explicit Integration Scheme

Time centered two-step central difference, the simplest explicit method, is conditionally stable and induces a time step stability condition based on the maximum frequency in the discretized system:

$$\Delta t \leq \Delta t_{\text{crit}} = \frac{2}{\omega_{\max}} = \frac{2}{\sqrt{\lambda_{\max}}} \tag{15.6}$$

where $\lambda_{\max}$ is the maximum eigenvalue of the finite element model. Noting that explicit expressions for the maximum eigenvalue of a system exist only in linear one-dimensional problems, the numerical solution of a general eigenvalue problem, accordingly to be solved, is often computationally prohibitive, therefore, alternative, more practical criteria are widely used. The most popular of the latter is based on the ratio of the length of the smallest element in the structure, during its dynamic response, to the speed of sound of that particular material:

$$\Delta t \leq \min\left(\frac{l_c^1}{c^1}, \frac{l_c^2}{c^2}, \frac{l_c^3}{c^3}, \ldots, \frac{l_c^n}{c^n}\right) \tag{15.7}$$

Excessive distortions during impact usually limit the time step to a number very small, to the extent that many thousands of cycles are needed, although no solution of the stiffness matrix is involved. With *orthotropic* materials, a good estimate for the speed of sound is based on the average stiffness coefficients in each of the three principal directions. In the present, the explicit two- and three-dimensional finite element codes LS-DYNA2D[6] and DYNA3D [4] are used. The discretized equations at the central time between t_{n-2} and t_n are t_{n-1} and are written as

$$[M]\{u\}_{n-1} + [C]\{u\}_{n-1} + [K]\{u\}_{n-1} = \{F\}_{n-1} \tag{15.8}$$

Approaching velocities and accelerations with respect to displacements with a two step recurrence relation only:

$$\{u\}_{n-1} \cong \frac{\{u\}_n - \{u\}_{n-2}}{2\Delta t} \quad \text{and} \quad \{u\}_{n-1} \cong \frac{\{u\}_n - 2\{u\}_{n-1} + \{u\}_{n-2}}{\Delta t^2} \tag{15.9}$$

one obtains:

$$\left[\frac{1}{\Delta t^2}[M] + \frac{1}{2\Delta T}[C]\right]\{u\}_n = \{F\}_{n-1} - \left[[K] + \frac{2}{\Delta t^2}[M]\right]\{u\}_{n-1} - \left[\frac{1}{\Delta t^2}[M] + \frac{1}{2\Delta t}[C]\right]\{u\}_{n-2} \tag{15.10}$$

One easily notes that, if matrices $[M]$ and $[C]$ are lumped, then the effective stiffness matrix in the LHS is independent of $[K]$, and, therefore, the

[6] J.O. Hallquist, *LS-DYNA2D User's manual*, Livermore, (1990).

equations are uncoupled and can be *explicitly* solved. It is estimated that the cost per increment is proportional to the number of elements. The central difference time integration scheme has no other constraints, apart from the critical time step, which translates to many iterations or large elements, wherever possible, depending on the bulk wave propagation velocity. Implicit integration schemes are also used in low and medium velocity impact problems, being not so highly nonlinear. The only drawback lies with cases, where the extra number of degrees of freedom, required to describe the contact regions, give rise to an even greater bandwidth, which, squared, leads to excessive computer cost. One should finally note, that explicit schemes perform, by default, dynamic analysis, since one is seeking for the fulfillment of the dynamic equilibrium:

$$[M]\{u\} = \{F\} - \{I\} \tag{15.11}$$

whereas, in the implicit procedures, only static equilibrium is required: $\{F\} = \{I\}$, where $\{F\}$, $\{I\}$ are the external forces applied and the internal element forces respectively.

15.5 Contact-Impact Algorithm

This section does not intend to elucidate the deep and rapid progress of the contact-impact algorithms, evolved within the last twenty years by the computational mechanics community, although the origins of their development is mainly coming from a handful of institutes, e.g., LLNL, UCB, MIT and University of Texas in the USA and recently University of Linköping, Sweden. The purpose of the present is an abstract description of the finite element procedure used in contact-impact problems, in order to be linked smoothly with the presentation of the results and conclusions drawn from the analysis of the impact response of Achilles' shield. Several survey articles as well as new formulations accounting for frictional phenomena, where the tangential component of motion is important, exist in the literature. Interested readers may consult the works of Hallquist, Goudreau and Benson,[7] and Hughes, Taylor, et al.[8] Contact boundary conditions can be described by the fact that

[7] J.O. Hallquist, G.L. Goudreau and D.J. Benson, Sliding interfaces with contact-impact in large scale Langrangian computations, *Comp. Mech. Appl. Mech. Engng.*, vol. 51, 1985, 107–137.

[8] T.J.R. Hughes, R.L. Taylor, J.L. Sackman, A. Curnier and W. Kanokhululchai, A finite element method for a class of contact-impact problems, *Comp. Mech. Appl. Mech. Engng.*, vol. 8, 1976, 249–276.

no boundary material particle belonging to a boundary Γ can penetrate the interior of the contact bodies Ω at any time, that is:

$$\Gamma \cap \Omega = \emptyset \tag{15.12}$$

where $\emptyset$ denotes the null set. Alternatively, the penetration between a boundary hitting and target particle can be described by the notion of *depth* for each boundary particle belonging to the global body remainder $\overline{\Omega}$. As abstractly posed in the introduction of this chapter, the best suited ways to handle the variational inequalities are via transformation to equalities and fulfillment of the constraint conditions through penalty or Lagrange multipliers techniques. The drawback for the former is that contact conditions are approximately satisfied, while for the latter, the increased number of equations that must be solved. A hybrid approach was suggested by Simo, Wriggers and Taylor,[9] called *perturbated Lagrangian formulation*, which includes both Lagrangian and penalty techniques. The latter is briefly outlined in the sequence. In the penalty method, the modified functional is obtained by adding a quadratic term associated with the constraint:

$$P(u,v) = I(u,v) + \frac{\gamma}{2}\int_{\alpha}^{\beta} [G(u,u',v,v')]^2 d\chi \tag{15.13}$$

where γ is the penalty parameter. As in Lagrange multiplier method, Euler's are obtained by setting the first variation of the modified functional equal to zero. With colliding bodies, the modified part of the functional assumes the form:

$$\gamma \int_c u_n^2 dc \tag{15.14}$$

where u_n is the normal displacement on the contact boundary. The physical interpretation of the penalty function is a stiffness factor for the master segment, given in terms of bulk modulus K, volume V, face area A and a scale factor for the interface stiffness:

$$k_i = \frac{f_{si} K_i A_i^2}{V_i} \tag{15.15}$$

More on penalty methods can be found in J.O. Hallquist et al. [5] and in Reddy.[10]

[9] J.C. Simo, P. Wriggers and R.L. Taylor, A perturbed Langrangian formulation for the finite element solution of contact problems, *Comp. Mech. Appl. Mech. Engng.*, vol. 50, 1985, 163–180.

[10] J.N. Reddy, *Energy and Variational Methods in Applied Mechanics*, John Wiley & Sons, New York, 1984.

Contact algorithms are devoted to searching contacting nodes and calculating the nodal force vector of the colliding surfaces. The search part of the algorithm finds, for each slave node, its nearest point on the master surface. Special algorithms are employed, when the slave node is located at the intersection of two or more master segments. The issue of primary importance in any contact problem is the calculation of nodal force vector, when two or more bodies collide. Sliding interface is the numerical tool to overcome the contact problem when two bodies collide. Most sliding techniques are based on the decomposition of the velocity and acceleration into normal and tangential components. Motions in the normal direction are continuous when materials are in contact. When $\vartheta b^1 \cap \vartheta b^2 \neq 0$ at any time, the constraints are imposed to prevent penetration. Each slave node is checked for penetration through the master surface, which is defined as:

$$\mathbf{I} = n_i \cdot [t - r(\xi_c, \eta_c)] < 0 \tag{15.16}$$

where $n_i = n_i(\xi_c, \eta_c)$ is normal to the master segment at the contact point, t is the position vector drawn to slave node and $r(\xi_c, \eta_c)$ the position vector at the contact point. If penetration occurs, an interface force is applied with a magnitude proportional to the amount of penetration.

15.5.1 Elastic-Plastic Constitutive Equations

As mentioned in the introduction, the choice of the constitutive equation is somehow a matter of experience in impact problems, since the validity of the choice should be reconsidered after careful examination of the preliminary results. Therefore, an elastic-plastic constitutive equation with hardening is chosen for all materials involved in the analysis. If the plastic strains resulting exceed unity, thermal softening should be considered, since in metals it is conceivable to have such highly-localized superplasticity, yet accompanied with the production of high temperatures. Steinberg and Guinan[11] material model has been used, accounting for temperature-dependent shear modulus and yield strength. The Gruneisen equation of state was used to describe the dilatational behavior of all materials.

[11] D.J. Steinberg and M.W. Guinan, A High Strain Rate Constitutive Model for Metals. University of California, Lawrence Livermore National Laboratory, Report, UCRL-80465, 1978.

15.5.2 Friction Model

As stated earlier, the contact algorithm prevents penetration of the slave nodes to the master surface using the decomposition of the velocity and acceleration into normal and tangential components. Thus, continuity in the normal direction is imposed using interfacial spring-force elements, while sliding occurs in the tangential direction. During sliding, friction forces develop, based on a Coulomb formulation. In each integration step, a frictional yield force is calculated using the normal force $\underline{fn}$ and the coefficient of friction μ_s according to

$$Fg = \mu|fn|$$

Then the active interface force is calculated based on interface stiffness and relative displacement between slave modes and master surface. In the sequence, a check for the interface yield follows.

An exponential interpolation function smoothens transition between static μ_s and dynamic μ_i coefficients of friction, where V is relative velocity between slave mode and master segment:

$$\mu = \mu_t + (\mu_c - \mu_i) \cdot e^{-c|V|}$$

where c is a decay constant.

The interface shear stress developing is also calculated.

Chapter 16
The Shield of Achilles

As described in Chapter 4, Achilles, succumbing to the pleas of the Achaeans, facing defeat in his absence, sent his friend Patroclus to fight in his stead, wearing Achilles' own panoply, to deceive the Trojans and make them think that Achilles himself had returned to battle. However, Patroclus was killed by Hector, and Achilles lost, besides his friend, his panoply as well. So he was unarmed, when, enraged, he truly wanted to return to war. Thetis, his mother, visited Hephaestus and asked him to fashion a panoply for her son. The visit of Thetis to Hephaestus has already been accounted for in Chapter 12, when referring to that miraculous forge. In the following paintings, one can find a few tokens of the inspiration caused to artists of all times by this event (Figures 16.1, 16.2, 16.3).

In Book 8 of the Iliad, the construction of Achilles' panoply by Hephaestus is described. In particular, in verses 474–482, immediately after the reference to Hephaestus "bellows" (Chapter 12, the purely technical account of the manufacture of Achilles' shield is given:

χαλκὸν δ' ἐν πυρὶ βαλλεν ἀτειρέα κασσίτερον τὲ
καὶ χρυσὸν τιμῆντα καὶ ἄργυρον αὐτὰρ ἔπειτα
θῆκεν ἐν ἀκμοθέτῳ μέγαν ἄκμονα, γέντο δὲ χειρὶ
ῥαιστῆρα κρατερήν, ἑτέρηφι δὲ γέντο πυράγρην.
ποίει δὲ πρώτιστα σάκος μέγα τε στιβαρὸν τε
πάντοσε δαιδάλλων, περὶ δ' ἄντυγα βάλλε φαεινὴν
τρίπλακα μαρμαρέην, ἐκ δ' ἀργύρεον τελαμῶνα.
πέντε δ' ἄρ αὐτοῦ ἔσαν σάκεος πτύχες· αὐτὰρ ἐν αὐτῷ
ποίει δαίδαλα πολλὰ ἰδίῃσι παρπίδεσσιν.

[Hephaestus ...] threw tough copper into the fire, and tin, with silver and gold; he set his great anvil on its block, and with one hand grasped his mighty hammer while he took the tongs in the other. First he shaped the

S.A. Paipetis, *The Unknown Technology in Homer*, History of Mechanism and Machine Science 9,
DOI 10.1007/978-90-481-2514-2_16, © Springer Science + Business Media B.V. 2010

(a)

(b)

Figure 16.1 Sir James Thornhill (1675–1734): (a) Three Studies for Thetis in the Forge of Vulcan Watching the Making of Achilles' Armour, (b) Thetis Accepting the Shield of Achilles from Vulcan (ca. 1710), Tate Gallery, London (reproduced by permission).

Figure 16.2 Anton van Dyck (1599–1641): Hephaestus delivering Achilles' armour to Thetis.

Figure 16.3 Hephaestus and the Cyclops forging Achilles' shield, Roman bas-relief.

shield so great and strong, adorning it all over and binding it round with a gleaming circuit in three layers; and the baldric was made of silver. He made the shield in five thicknesses, and with many a wonder did his cunning hand enrich it.

(Il. 18.474–482)

This description corresponds to a *multi-layered structure*, consisting of five successive metal laminates with very different mechanical properties. In particular, the shield consists of two external laminates of hard bronze, two internal ones of tin and a central laminate of pure, i.e., soft, gold. This structure exhibits maximum resistance to penetration, as proved by a complete numerical simulation of large-strain elasto-plastic behaviour when impacted by the tip of a piercing element, such as the tip of a spear or an arrow. The analysis and parametric study of the problem, as shown in the sequence, using modern continuum mechanics theories and the numerical tools of Chapter 15, have most accurately confirmed the ensuing Homeric descriptions, pertaining to the battle behaviour of the shield most accurately and revealed important elements of advanced technology in Mycenaean times, disguised as the miraculous power of gods.

Mycenaean shields appear in various shapes, mainly as eight-shaped (Figure 16.4), circular (Figures 16.5 and 16.6) or full-body shields. Note the representations of ghastly beings on the shields, aiming to intimidate the enemy.

The Homeric description classifies Achilles' shield as circular (Figure 16.6). Numerous reconstructions of it have been made (Figure 16.7), especially reproducing its decorations, which are accounted for in detail in Il. 18.483–608. Out of those, the excellent reconstruction in gold by Philip Rundell (1821) on the coronation of King George IV, is distinguished (Figure 16.8).

The battle behaviour of the shield of Achilles is described in the Iliad on three distinct occasions. (a) The duel between Achilles and Aeneias:

Ἦ ῥὰ καὶ ἐν δεινῷ ἤάκει ἤλασεν ὄβριμον ἔγχος
σμερδαλέῳ· μέγα δ' ἀμφὶ σάκος μύκε δουρὸς ἀκωκῇ.
Πηλεΐδης δὲ σάκος μὲν ἀπὸ ἕω χειρὶ παχείῃ
ἔσχετο ταρβήσας· φάτο γαρ δολιχόσκιον ἔγχος
ῥέα διελεύσεσθαι μεγαλήτορος Αἰνείαο
νήπιος, οὐδ' ἐνόησε κατὰ φρένα καὶ κατὰ θυμὸν
ὡς οὐ ῥηΐδι ἐστὶ θεῶν ἐρικυδέα δῶρα
ἀνδράσι γὲ θνητοῖσι δαμήμεναι οὐδ' ὑποείκειν.
οὐδὲ τότ' Αἰνείαο δαΐφρονος ὄβριμον ἔγχος
ρῆξε σάκος· χρυσὸς γὰρ ἐρήκακε, δῶρα θεοῖο·
ἀλλὰ δύω μὲν ἔλασσε διὰ πτύχας, αἱ δ' ἄρ' ἔτι τρεῖς
ἦσαν, ἐπὶ πέντε πτύχας ἤλασε κυλλοποδίων,
τὰς δυὸ χαλκείας, δύο δ' ἔνδοθι κασσιτέροιο,
τὴν δὲ μίαν χρυσῆν, τῇ ρ' ἔσχετο μείλινον ἔγχος.

… The son of Peleus
held the shield before him with his strong hand, and he was afraid, for he

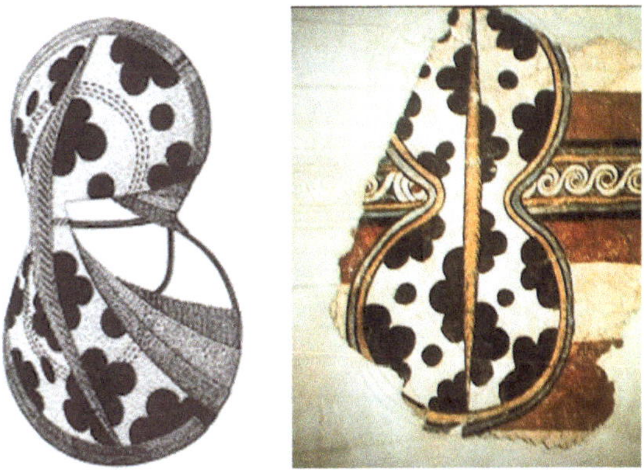

Figure 16.4 Sketch of a eight-shaped shield and its representation on a mosaic.

Figure 16.5 Typical geometry of a circular Mycenaean shield (12th century BCE).

Figure 16.6 Typical decorated shields of the Homeric era.

Figure 16.7 Two representations of Achilles' shield based on the Homeric description.

Figure 16.8 Shield of Achilles: Reproduction in gold by Philip Rundell. Reproduced by permission, The Royal Collection © 2009 Her Majesty Queen Elizabeth II.

deemed that Aeneas's spear would go through it quite easily, not reflecting that the god's glorious gifts were little likely to yield before the blows of mortal men; and indeed Aeneas's spear did not pierce the shield, for the layer of gold, gift of the god, stayed the point. It went through two layers, but the god had made the shield in five, two of bronze, the two innermost ones of tin, and one of gold; it was in this that the spear was stayed ...

(Il. 20.259–270)

(b) The duel between Achilles and Hector:

ἦ ῥά, καὶ ἀμπεπαλὼν προΐει δολιχόσκιον ἔγχος,
καὶ βάλε Πηλεΐδαο μέσον σάκος οὐδ' ἀφάμαρτε·
τῆλε δ' ἀπεπλάγχθη σάκεος δόρυ·
... He poised his spear as he spoke and hurled it. His aim was true for he hit the middle of Achilles' shield, but the spear rebounded from it, and did not pierce it ...

(Il. 22.289–292)

(c) The duel between Achilles and Asteropaeus, the Trojan hero:

ὣς φάτ' ἀπειλήσας, ὁ δ' ἀνέσχετο δῖος 'Αχιλλεὺς
Πηλιάδα μελίην· ὁ δ' ἁμαρτῇ δούρασιν ἀμφὶς
ἥρως 'Ἀστεροπαῖος, ἐπεὶ περιδέξιος ἦεν.
καί ῥ' ἑτέρῳ μὲν δουρὶ σάκος βάλεν, οὐδὲ διὰ πρὸ
ῥῆξε σάκος· χρυσὸς γὰρ ἐρύκακε δῶρα θεοῖο·

Thus did he defy him, and Achilles raised his spear of Pelian ash. Asteropaeus failed with both his spears, for he could use both hands alike; with the one spear he struck Achilles' shield, but did not pierce it, for the layer of gold, gift of the god, stayed the point;

(Il. 21.161–164)

These references are highly enlightening and provide the basis for the numerical simulation of the weapons, i.e., of a spear with a hard bronze tip impacting the shield. The exact geometry of the shield and the spear are not known, but based on archaeological findings and schematic or painted representations, fundamental information for the analysis has been obtained. The geometry of the shield was presented already, while a number of typical bronze spearheads appear in Figure 16.9.

16.1 Numerical Analysis and Results[1]

Assume that each of the five laminates of Achilles' shield is 1.5 mm thick, i.e., the total thickness of the shield is 7.5 mm. Considering a spear hitting

[1] The numerical analysis was performed at the Applied Mechanics Laboratory of the University of Patras (Professor S.A. Paipetis, Director) by a research team under Professor V. Kostopoulos.

Figure 16.9 Characteristic decorated spear-heads used in the Trojan War.

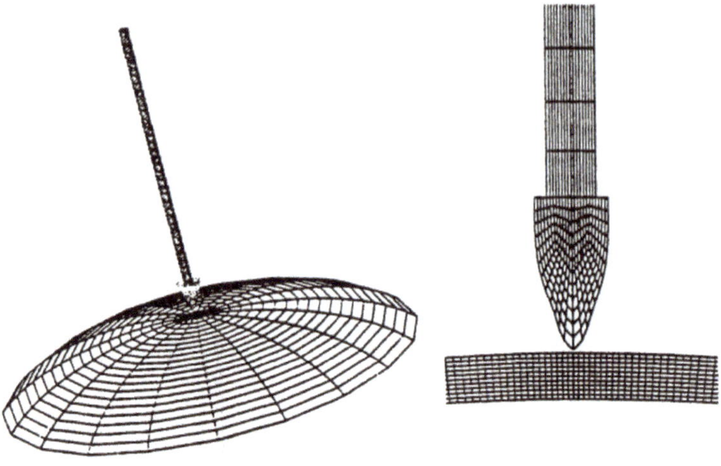

Figure 16.10 Finite Element discretization of the spear-shield system: The complete three-dimensional model (left) and a detail of spearhead (pike) and shield contact point (right).

it at its centre in a direction normal to the surface, an axisymmetric problem results. In Figure 16.10 details of the shield cross section is given along with its three-dimensional discretization. The form of the generatrix of the shield cross section is elliptic with semi-axes 300 and 120 mm respectively. This

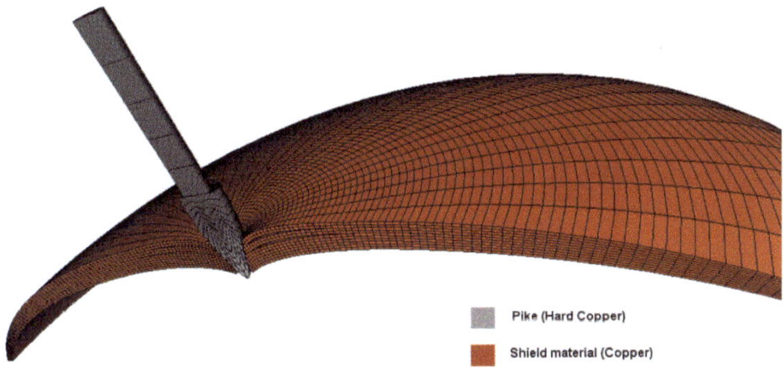

Figure 16.11 Impact behaviour of the spear-shield system, with all shield laminates being of bronze, at 4 msec time after contact, by which the spear penetrates the shield.

leads to a total mass of 16.75 kg. The spear is taken as a rod 2.2m long with 16 mm diameter.

The form of spear tip was based on vase paintings and archaeological findings. As often described in the Iliad, bronze spearheads were fastened on long wooden rods with circular cross section. The dimensions and materials of the spear correspond to a total mass of 3.25 kg. As stated, the shield was studied as a multi-layered structure, and the way in which the laminates were interconnected was a key-point.

Furthermore, it was assumed that the edges of the front laminate were turned backwards, encasing the whole structure. In addition, a number of bolts kept the laminates together across the thickness, securing good mechanical cooperation. The number and arrangement of the bolts is very important for the impact response of the shield and stress distribution.

It was further assumed that the bolts were arranged circumferentially on a circle at 120 mm distance from the shield axis of symmetry. Such an arrangement allows the laminates to slide in relation to each other.

The spear velocity was taken equal to 20 m/sec. Two specific cases were studied. In the first, all shield laminates consisted of hard bronze (or "copper"), while, in the second, the true (according to Homer) formation of the shield is considered, e.g., two external bronze laminates, two internal tin laminates and a central laminate of gold. The spearhead consists of hard bronze.

The properties of the materials used were given in Chapter 15. In Figure 16.11 the response of the shield when all laminates were of bronze is presented. In this case, the spear penetrates the shield. In Figures 16.12 and 16.13 the respective displacement and velocity diagrams of spear tip and the

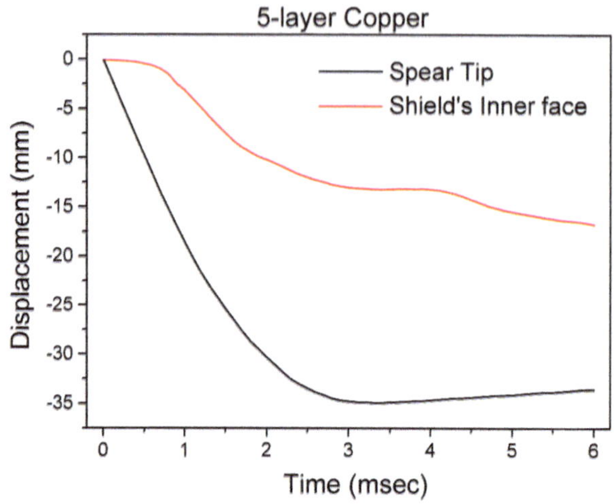

Figure 16.12 Displacements diagrams of spear tip and of the central point of the shield plotted against time, when all shield's laminates are of bronze (5 bronze – or "copper" – laminates).

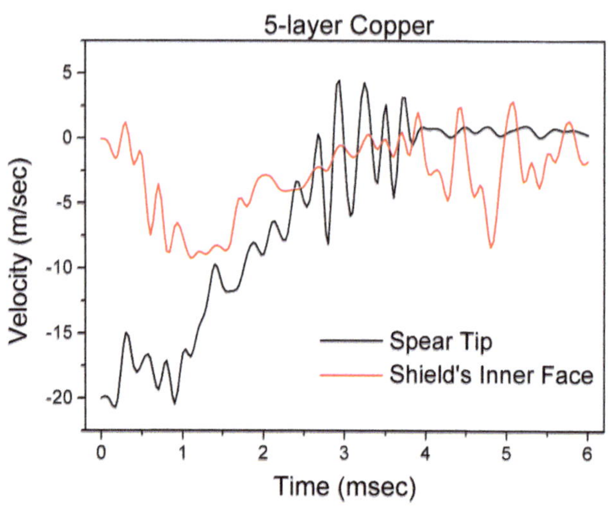

Figure 16.13 Velocity diagram of spear tip and of the central point of the shield plotted against time, when all shield's laminates are of bronze (5 bronze – or "copper" – laminates).

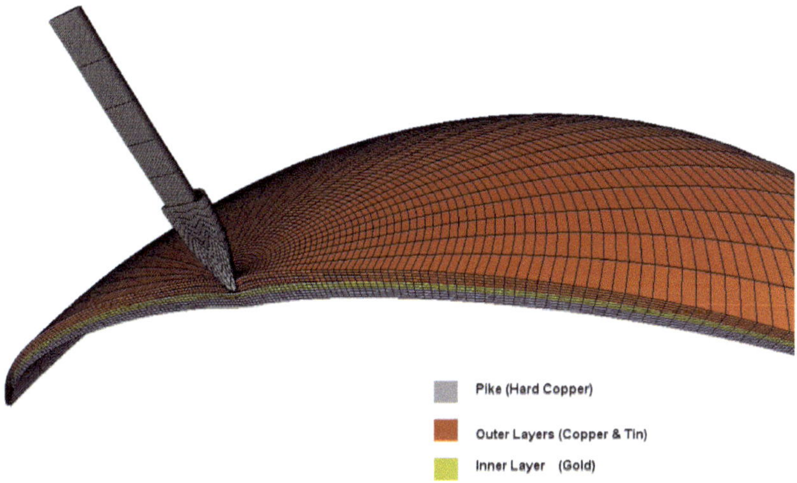

Figure 16.14 Impact behaviour of spear-shield system based on the real properties of the individual shield laminates (bronze-tin-gold) at 1.5 msec time, in which case the spear penetrates the two external laminates, i.e., bronze and tin, and is stopped by the gold laminate.

central point of the shield are plotted against time. Penetration of the shield is complete after 4 ms, while the change in slope of the velocity diagram of the spear tip corresponds to the ensuing penetration into the bronze laminates. At the end of the penetration stage, the spear has almost zero velocity, while the shield displacement reaches 35 mm.

Figure 16.14 presents the response of the simulated Achilles' shield. In this case, the spear tip stopped at the gold laminate, having pierced the two external bronze and tin laminates. This occurs within 1.5 msec, while, in the sequence, the spear rebounded. Figure 16.15 presents the displacement and velocity diagrams of the spear tip and the shield's central point respectively plotted against time.

In conclusion, one can note that: (a) the shield consisted of a layering of laminates with very different mechanical properties, of which the strongest was hard bronze, (b) the combination of these materials, in the way described in the Iliad, prevented the spear from penetrating the shield, since, on one occasion, it was repulsed and rebounded, while, on the other, it only pierced the first two laminates (bronze-tin) and subsequently it was stopped by the gold laminate, and (c) if all laminates of the structure consisted of bronze, under the same impact conditions, the shield would be penetrated.

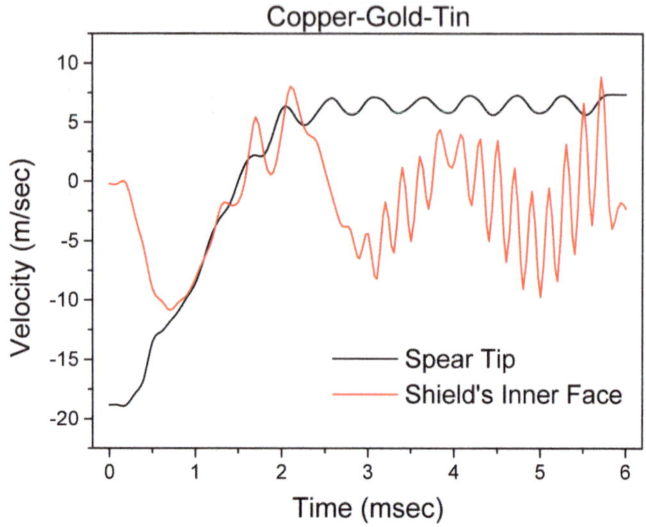

Figure 16.15 Velocity diagrams for spear tip and the central point of the shield by using the "real" materials (bronze-tin-gold) of Achilles' shield.

This fact is due to the complete difference in behaviour of the materials under static and dynamic loading. The object of a defensive device, like a shield, is to "destroy" (in fact to scatter) the kinetic energy of the projectile, by converting it into heat, and not just to sustain a large static load. This ability is not possessed by hard bronze itself, exhibiting very small deformations in relation to the rest of the materials and also possessing very low damping properties. On the contrary, tin but, mainly, pure (i.e., soft) gold, while undergoing plastic deformation, causes attenuation of the motion by dissipating the kinetic energy of the spear. Moreover, the laminated structure contributes by a certain amount of damping due to friction between the laminates, which, however, is not the main energy dissipation mechanism.

Based on these remarks, the hypothesis that the shield manufacturer possessed deep knowledge of the dynamic mechanical properties of laminated composite structures, essentially nearly modern structural elements, is fully confirmed.

Chapter 17
The Shield of Ajax

Ajax, son of Telamon, King of Salamis, and, according to Homer, a man of enormous stature and a colossal body, was inferior in strength and bravery only to Achilles (Figure 17.1).

Ajax fought against Hector and, by the help of Athena, he saved Achilles' body from the hands of the Trojans (Figure 17.2). Ajax lost in a contest with Odysseus for the possession of Achilles' panoply, and committed suicide.

Figure 17.1 Henri Serrur, Death of Ajax (1820), Musée des Beaux Arts, Lille, France.

S.A. Paipetis, *The Unknown Technology in Homer*, History of Mechanism and Machine Science 9, DOI 10.1007/978-90-481-2514-2_17, © Springer Science + Business Media B.V. 2010

(a) (b)

Figure 17.2 Ajax with the body of Achilles. (a) Greek amphora of the 5th century BCE, (b) seal-ring of engraved cornelian, Etruria, 5th cent. BCE, The State Hermitage Museum, St. Petersburg (Photograph © The State Hermitage Museum, reproduced by permission).

According to a later version, his disappointment led him to madness. When he came around, he committed suicide, by throwing himself on a sword that had been a present from Hector (Figure 17.3).

Ajax Telamonius was the patron hero of Salamis Island, where a temple and a statue were erected in his honour, while *Aianteia* was an annual festivity. A town in Salamis is still called *Aianteion.*

Ajax participated to the Trojan War and the siege of Troy. He also possessed an extraordinary shield, described with great clarity in the Iliad, and with sufficient details to perform a numerical simulation. Since the materials specified are not as expensive, as compared to the gold content of the shield of Achilles, and the construction much simpler, it was possible to produce specimens and perform an experimental study, to confirm analytical results and, in particular, to investigate the mechanism of energy dissipation of the impactor, which is radically different than the one of Achilles' shield. The Homeric description follows:

Αἴας δ’ ἐγγεῖθεν ἦλθε φέρων σάκος ἠΰτε πύργον
χάλκεον ἑπταβόειον, ὅ οἱ Τυχίος κάμε τεύχων

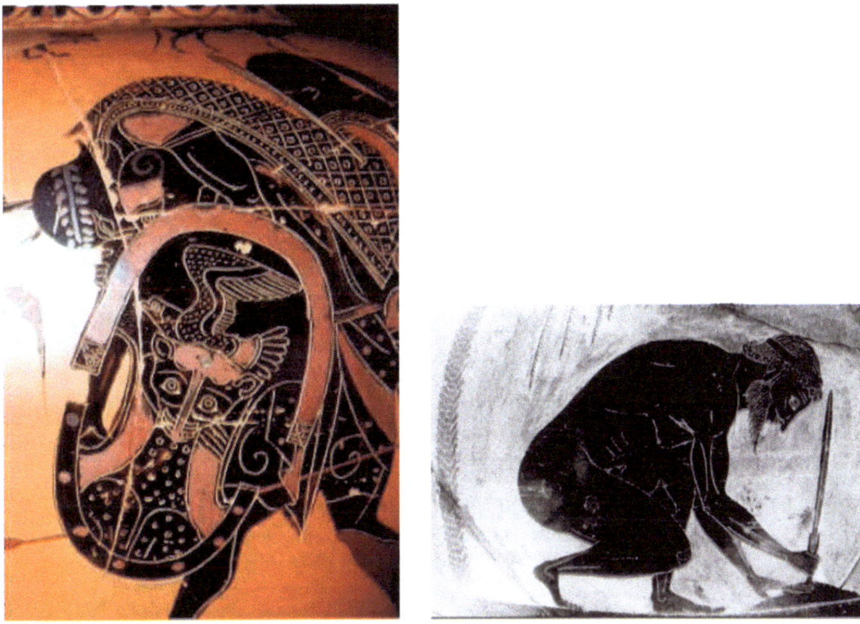

Figure 17.3 Death of Ajax: Black-painted attic amphoras of the 5th century BCE with the shield (left) and suicide (540–530 BC), work of Execias (550–525 BCE).

σκυτοτόμων ὄχ' ἄριστος Ὕλῃ ἔνι οἰκία ναίων,
ὅς οἱ ἐποίησεν σάκος αἰόλον ἑπταβόειον
ταύρων ζατρεφέων, ἐπὶ δ' ὄγδοον ἤλασε χαλκόν.

Ajax came up bearing his shield in front of him like a wall – a shield of bronze with seven folds of oxhide – the work of Tychius, who lived in Hyle and was by far the best worker in leather. He had made it with the hides of seven full-fed bulls, and over these he had set an eighth layer of bronze.
(Il. 7.219–223)

The battle behaviour of Ajax's shield during his duel with Hector is described in the sequence in the same book:

ἦ ῥα καὶ ἀμπεμπαλῶν προΐει δολιχόσκιον ἔγχος,
καὶ βάλεν Αἴαντος δεινὸν σάκος ἑπταβόειον
ἀκρότατον κατὰ χαλκόν, ὅς ὄγδοος ἦεν ἐπ' αὐτῷ.
ἕξ δὲ διὰ πτύχας ἦλθε δαΐζων χαλκὸς ἀτειρής,
ἐν τῇ δ' ἑβδομάτῃ ῥινῷ σχέτο· δεύτερος αὖτε
Αἴας διογενὴς προΐει δολιχόσκιον ἔγχος
καὶ βάλε Πριαμίδαο κατ' ἀσπίδα πάντοσ' ἐΐσην.
διὰ μὲν ἀσπίδος ἦλθε φαεινῆς ὄβριμον ἔγχος,
καὶ διὰ θώρακος πολυδαιδάλου ἠρήρειστο·

ἀντικρὺ δέ παραὶ λαπάρην διάμησε χιτῶνα
ἔγχος· ὁ δ' ἐκλίνθη καὶ ἀλεύατο κῆρα μελαίναν.
τὼ δ' ἐκσπασσαμένω δολίχ' ἔγχεα χερσὶν ἅμ' ἄμφω
σὺν ῥ' ἔπεσον λείουσιν ἐοικότες ὠμοφάγοισιν
ἢ συσὶ κάπροισιν, τῶν τε σθένος οὐκ ἀλαπαδνόν.
Πριαμίδης μὲν ἔπειτα μέσον σάκος οὔτασε δουρί,
οὐδ' ἔρρηξεν χαλκός, ἀνεγνάμφθη δέ οἱ αἰχμή.

He poised his spear as he spoke, and hurled it from him. It struck the sevenfold shield in its outermost layer – the eighth, which was of bronze – and went through six of the layers but in the seventh hide it stayed. Then Ajax threw in his turn, and struck the round shield of the son of Priam. The terrible spear went through his gleaming shield, and pressed onward through his cuirass of cunning workmanship; it pierced the shirt against his side, but he swerved and thus saved his life. They then each of them drew out the spear from his shield, and fell on one another like savage lions or wild boars of great strength and endurance: the son of Priam struck the middle of Ajax's shield, but the bronze did not break, and the point of his dart was turned.
(Il. 7. 244–259)

Hence, the shield is also a multi-layered structure, consisting of eight consecutive laminates, e.g., of a front laminate of hard bronze and seven layers of calf's leather. As shown in the sequence, both numerically and experimentally, the Homeric descriptions of shield battle behaviour are confirmed with surprising accuracy and, once more, reveal elements of advanced scientific and technological knowledge.

17.1 Analysis of Results

In this section the impact of Hector's spear on Ajax's shield is analyzed. The same assumptions as with Achilles' shield were adopted, concerning shape and dimensions of spear and shield, and data from the modern javelin sport have been utilized.

It was assumed that the seven leather layers of the shield were of equal thickness, varying from 1 to 1.5 mm. With a circular shield and a spear hitting it at a direction normal to its surface, the problem is axisymmetric. The form of the generatrix of the shield cross section is elliptic with semi-axes 300 and 120 mm. Details of shield cross section and the spear head, as well as their three-dimensional discretization, were already given in Figures 16.10 and 16.11, concerning results for layers of bronze only (as in the analysis of Achilles' shield; these figures are repeated). Also, in Figure 17.4, as an

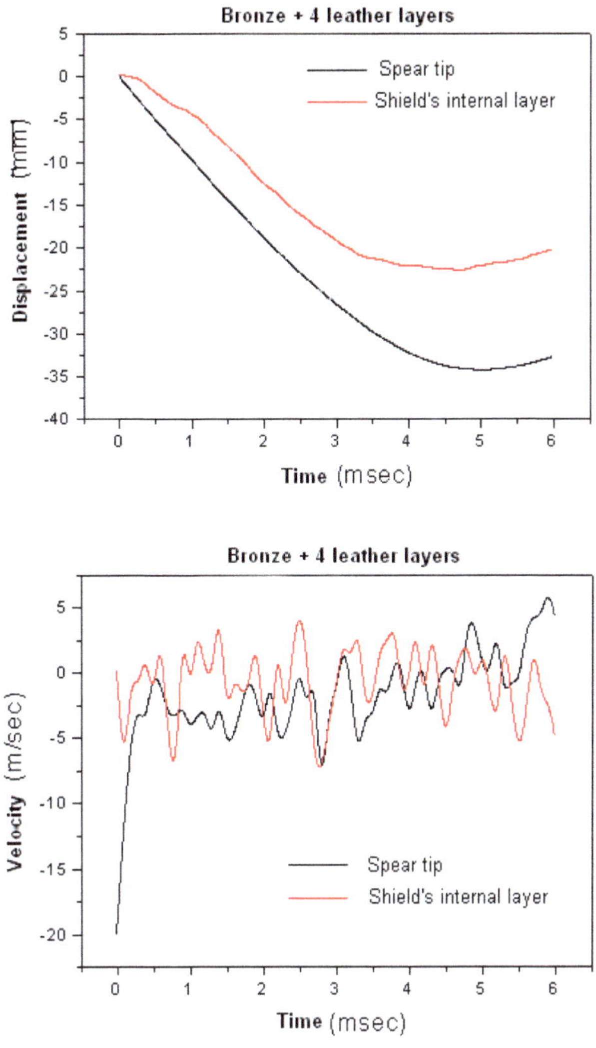

Figure 17.4 Penetration diagram of a simulated shield, consisting of one bronze laminate and four leather layers.

example, the results for one bronze laminate and four leather layers are given. In the present analysis, as the main factor of spear motion attenuation, the friction between the leather layers was considered, which in this case is the main mechanism to convert its kinetic energy into heat.

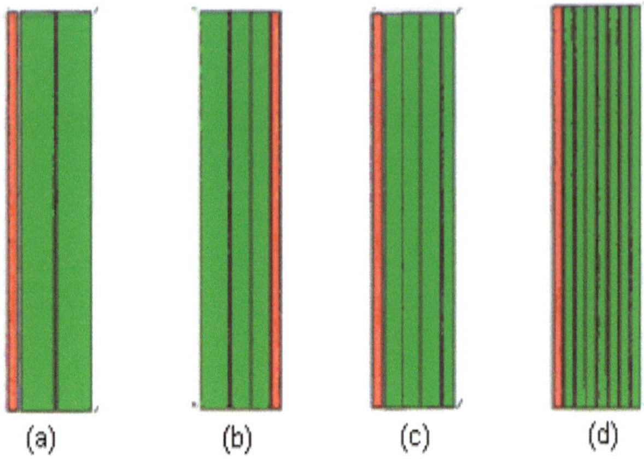

Figure 17.5 Several specimen forms: bronze laminate (red) is 1 mm thick, while total thickness for all leather layers (green) is 7 mm. There are in (a) 2 × 3.5 mm leather layers, in (b) 2 × 2 mm + 1 × 3 mm, in (c) 3 × 2 mm + 1 mm and in (d) 7 × 1 mm. Specimen dimensions are 140 × 140 mm. Only specimen (d) resisted penetration.

17.2 Experimental

To test the penetration resistance, a series of specimens were manufactured, having the same multi-layer structure consisting of one laminate of hard bronze and a number of layers of calf' leather. To investigate the mode of operation of the shield and of spear penetration mechanism, various combinations of number and thickness of layers were examined. In fact, by maintaining the total thickness of leather layers constant, e.g., 7 mm, different forms of specimens, presented in Figure 17.5, were experimentally tested. The latter were impacted by hard bronze projectiles launched by an air-gun.

The experimental setup appears in Figure 17.6 and consists of the air-gun, a specimen clamping jig and a system to measure projectile velocity. The air-gun has a compressed-air vessel and a barrel.[1] The vessel is equipped with an electric detonation valve and a pressure gauge. The air-gun accepts barrels of different diameters and length. In this case, the barrel was 2m long and had a diameter of 13.4 mm. Compressed-air was supplied by a central distribution line. The specimen clamping jig with part of the barrel appears in Figure 17.7.

[1] Designed by the late Professor Werner Goldsmith, University of California at Berkeley in 1981 during his sabbatical year in Patras.

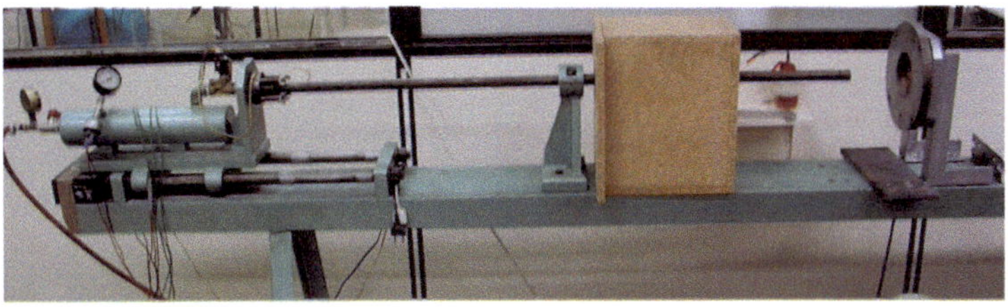

Figure 17.6 The air-gun.

Figure 17.7 Specimen clamping jig.

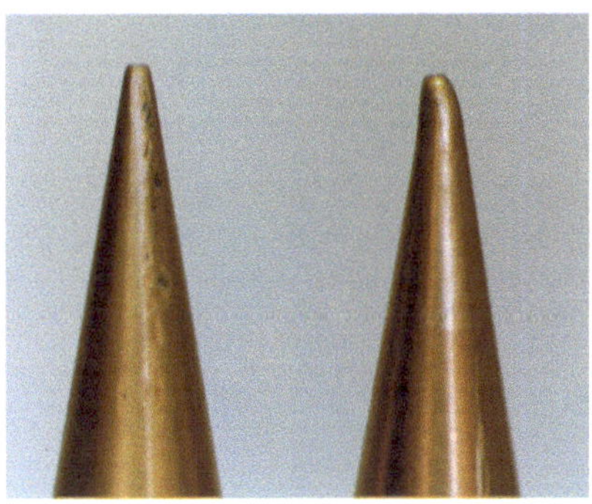

Figure 17.8 The projectile before and after impact.

The air-gun shoots cylindrical projectiles with a conical tip made of bronze harder than shield bronze (Figure 17.8). To measure the projectile velocity a split He-Ne laser beam and two photodiodes with their respective outputs fed into a CRT oscilloscope were used. The initial projectile velocity of a given mass projectile, again, corresponded to a kinetic energy equal to that of the javelin of year 2000 world champion, or 39.38 Joule.

The photograph of a specimen appears in Figure 17.9, and penetration hole on bronze surface in Figure 17.10.

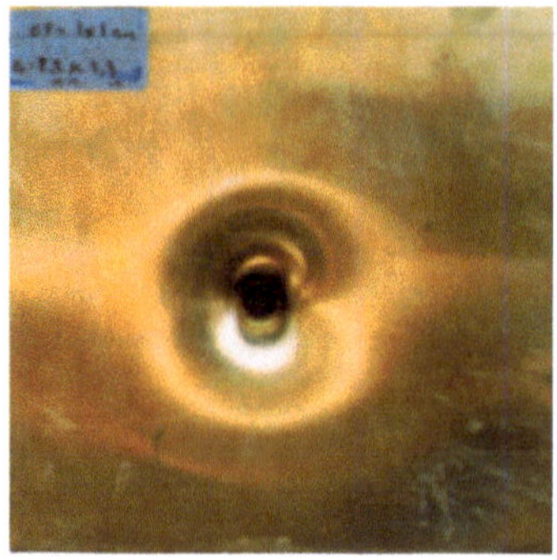

Figure 17.9 Penetration hole on bronze surface.

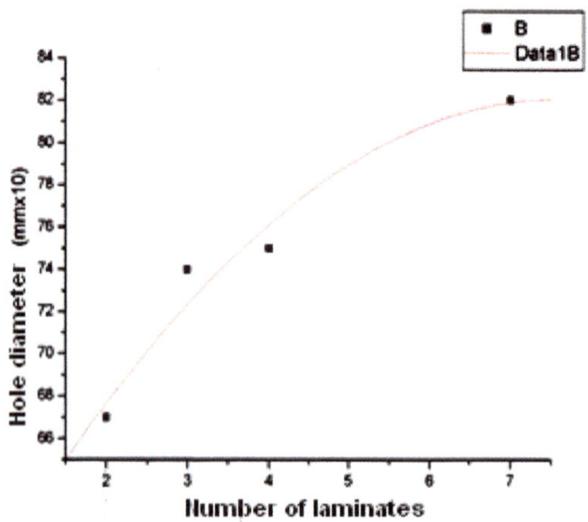

Figure 17.10 Penetration hole diameter on the surface of a bronze laminate against number of leather layers.

17.3 Discussion of Results and Conclusions

A measure of the projectile penetration into the respective specimen is the size of the hole created on the surface bronze laminate (Figure 17.10). The results for the various specimens appear in Figure 17.5. It is interesting that for the multi-layer shield with seven leather layers, i.e., Ajax's shield, penetration assumes a maximum value. However, this was the specimen that resisted penetration, and, in reality, was the last leather layer that eventually stopped the projectile. This agrees admirably with the Homeric description for its battle behaviour, stating that:

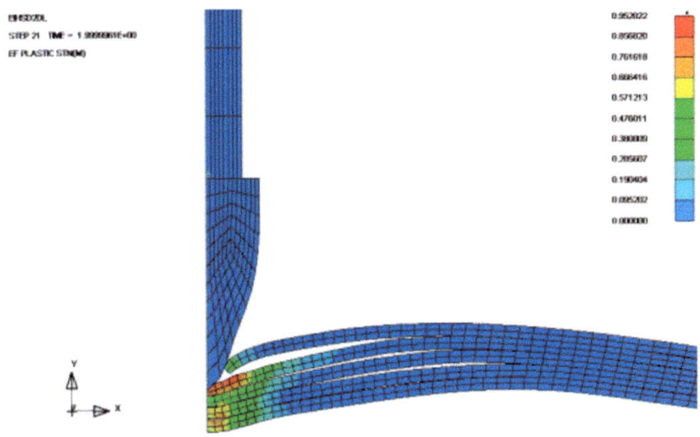

Figure 17.11 Animated representation of penetration of a model shield consisting of a bronze laminate and four leather layers.

> ['Εκτωρ]
> ... ἀμπεπαλὼν προΐει δολιχόσκιον ἔγχος,
> καὶ βάλεν Αἴαντος δεινὸν σάκος ἑπταβόειον
> ἀκρότατον κατὰ χαλκόν, ὃς ὄγδοος ἦεν ἐπ' αὐτῷ.
> ἓξ δὲ διὰ πτύχας ἦλθε δαΐζων χαλκὸς ἀτειρής,
> ἐν τῇ δ' ἑβδομάτῃ ῥινῷ σχέτο· ...
>
> [Hector] poised his spear as he spoke, and hurled it from him. It struck the sevenfold shield in its outermost layer – the eighth, which was of bronze – and went through six of the layers but in the seventh hide it stayed.
> (Il. 7.244–247)

The bending of the projectile tip after a failed impact, shown in Figure 17.8, is also mentioned, e.g.:

> Πριαμίδης μὲν ἔπειτα μέσον σάκος οὔτασε δουρί,
> οὐδ' ἔρρηξεν χαλκός, ἀνεγνάμφθη δέ οἱ αἰχμή.
>
> ... the son of Priam struck the middle of Ajax's shield, but the bronze did not break, and the point of his dart was turned.
> (Il. 7.258–259)

With the shield of Ajax, the kinetic energy of the projectile is absorbed by the friction between the layers of leather, acting efficiently for a number of layers equal to or greater than seven and with sufficient deformation of the layers. On the other hand, a polynomial interpolation in the results of Figure 17.11, shows that, for a greater number of leather layers, penetration tends to assume constant value. This proves that the structure described in the Iliad provides maximum resistance towards penetration, being also an optimum solution, since, for a greater number of layers, friction would be too strong and the structure too rigid to deform sufficiently and consume kinetic energy through friction. It is also noted that a small part of of the projectile kinetic energy is absorbed by the deformation of the projectile tip, mentioned in Il. 7.259 and appears in Figure 17.8.

Again, the hypothesis is confirmed, that the shield manufacturer (in this case, not Hephaestus, a god, but Tychius, a simple tanner), was fully aware of the dynamic mechanical behaviour of multi-layered composite structures.

Chapter 18
More Defensive Weapons

Because of their technological interest and the possibility of reconstructing and studying them on the basis of information given in various texts, we will give some further accounts of defensive weaponry in Homer and in antiquity in general. In the following, the shield of Heracles is presented, as described by Hesiod, as well as armours and helmets of the Mycenaean Era, and, finally, for comparison, a description of a Roman shield by Polybius.

18.1 The Shield of Heracles

An earlier reference to a shield with some structural details, not sufficient to reconstruct it, can be found in Hesiod concerning the shield of Heracles and his duel with Cygnus, son of Ares (Figure 18.1):

Χερσί γε μὴν σάκος εἷλε παναίολον, οὐδέ τις αὐτὸ
οὔτ' ἔρρηξε βαλὼν οὔτ' ἔθλασε, θαῦμα ἰδέσθαι.
πᾶν μὲν γὰρ κύκλῳ τιτάνῳ λευκῷ τ' ἐλέφαντι
ἠλέκτρῳ θ' ὑπολαμπὲς ἔην χρυσῷ τε φαεινῷ
[λαμπόμενον, κυάνου δὲ διὰ πτύχες ἠλήλαντο].
ἐν μέσσῳ δ' ἀδάμαντος ἔην Φόβος οὔ τι φατειός,
ἔμπαλιν ὄσσοισιν πυρὶ λαμπομένοισι δεδορκώς
τοῦ καὶ ὀδόντων μὲν πλῆτο στόμα λευκαθεόντων,
δεινῶν, ἀπλήτων, ἐπι δὲ βλοσυροῖο μετώπου
δεινὴ Ἔρις πεπότητο κορύσσουσα κλόνον ἀνδρῶν,
σχετλίη, ἥ· ῥα νόον τε καὶ ἐκ φρένας εἵλετο φωτῶν
οἵτινες ἀντιβίην πόλεμον Διὸς υἷι νέροιεν.
[τῶν καὶ ψυχαὶ μὲν χθόνα δύνουσ' Ἄιδος ἔσω
αὐτῶν, ὀστέα δὲ σφι περὶ ῥινοῖο σαπείσης
Σειρίου ἀζαλέοιο κελαινῇ πύθεται αἴῃ.]

S.A. Paipetis, *The Unknown Technology in Homer*, History of Mechanism and Machine Science 9,
DOI 10.1007/978-90-481-2514-2_18,

Figure 18.1 Heracles with the shield and Cygnus (left) and detail of the shield (right), Toledo Art Museum.

> In his hands he took his shield,[1] all glittering: no one ever broke it with a blow or crushed it. And a wonder it was to see; for its whole orb was a-shimmer with enamel and white ivory and electrum, and it glowed with shining gold; and there were zones of cyanus drawn upon it. In the centre was Fear worked in adamant, unspeakable, staring backwards with eyes that glowed with fire. His mouth was full of teeth in a white row, fearful and daunting, and upon his grim brow hovered frightful Strife who arrays the throng of men: pitiless she, for she took away the mind and senses of poor wretches who made war against the son of Zeus. Their souls passed beneath the earth and went down into the house of Hades; but their bones, when the skin is rotted about them, crumble away on the dark earth under parching Sirius.
>
> (Hesiod, Shield of Heracles, 139–153)

18.1.1 Cyanus

Cyanus was a glass paste of deep blue color: the "zones" were concentric circular bands, on which the scenes described by the poet were depicted. The picture of Fear (l. 44) occupied the centre of the shield, and Oceanus (l. 314) enclosed the whole. Here, one discovers the "aggressive" aspect of the shield, i.e., to intimidate the opponent by its terrifying representations.

[1] This shield was made by Hephaestus, and it was a wondrous work even for the king of gods.

Cyanus as structural material is repeatedly mentioned in the Iliad, in particular, as one of the materials of the breastplate of Atreides'[2] armor, a guest-gift of Cinyras of Cyprus:[3]

τοῦ δ' ἤτοι δέκα οἶμοι ἔσαν μέλανος κυάνοιο,
δώδεκα δὲ χρυσοῖο καὶ εἴκοσι κασσιτέροιο·
κυάνεοι δὲ δράκοντες ὀρωρέχατο προτὶ δειρὴν
τρεῖς ἑκάτερθ' ἴρισσιν ἐοικότες, ἅς τε Κρονίων
ἐν νέφεϊ στήριξε, τέρας μερόπων ἀνθρώπων.

It had[4] ten courses of dark cyanus, twelve of gold, and ten of tin. There were serpents of cyanus that reared themselves up towards the neck, three upon either side, like the rainbows which the son of Saturn has set in heaven as a sign to mortal men.

(Il. 11.24–28)

In the same Book, the reception of Patroclus and Machaon in Nestor's tent is given:

τοῖσι δὲ τεῦχε κυκειῶ ἐϋπλόκαμος 'Εκαμήδη,
τὴν ἄρετ' ἐκ Τενέδοιο γέρων, ὅτε πέρσεν 'Αχιλλεύς,
θυγατέρ' 'Αρσινόου μεγαλήτορος, ἥν οἱ 'Αχαιοί
ἔξελον οὕνεκα βουλῇ ἀριστενεσκεν ἁπάντων,
ἥ σφωϊν πρῶτον μὲν ἐπιπροΐηλε τράπεζαν
καλὴν κυανόπεζαν ἐΰξοον, αὐτὰρ επ' αὐτῆς
χάλκειον κάνεον, ἐπὶ δὲ κρόμυον ποτῷ ὄψον.

Fair Hecamede, whom Nestor had had awarded to him from Tenedos when Achilles took it, mixed them a mess; she was daughter of wise Arsinous, and the Achaeans had given her to Nestor because he excelled all of them in counsel. First she set for them a fair and well-made table that had feet of cyanus; on it there was a vessel of bronze and an onion to give relish to the drink, with honey and cakes of barley-meal.

(Il. 11.624–630)

---

[2] I.e., of Agamemnon.

[3] Verse 20. Cinyras, a hero of Cyprus, brought the worship of Aphrodite from Syria to Paphos. Many people, among them Tyrteus, compare him to Midas because of his innumerable wealth. This is the only reference to Cyprus in the Iliad, while in the Odyssey the island is mentioned more often.

[4] I.e., the breastplate.

18.1.2 Electrus

According to Zeggelis, electrus is not the well-known resin named ember, but a certain metal, probably a gold and silver alloy, a fact confirmed by several ancient historians.[5]

18.1.3 Ivory

Ivory, the material of elephant tusks, is a variation of dentine. It exhibits beauty, great durability and suitability for carving or sculpting with woodworking tools. It cannot be damaged, destroyed or burnt, and is hardly affected when immersed in water. Its properties resemble to those of hard wood, it has high density and can be easily polished. Since it is available in large size, it is suitable for the manufacture of defensive weapons, e.g., shields, due both to its toughness and the possibility of carving decorative depictions. This is not the case with the tusks of hippopotamus, sea-horse, unicorn whale, wild boar, and other species, which may be classified as ivory, but they have a small size, and respectively low commercial value. However, wild boar tusks were used to manufacture military helmets in antiquity, as it will be seen in the sequence.

Since very old times, ivory was considered a luxury material due to its subtle texture, light cream color and mild lustre. It has been used in all ancient civilizations of Egypt, China, Japan etc. but also in more recent ones, e.g., Roman, Byzantine, Renaissance, Baroque and present times.

Originating from the 16th century BCE, small carved ivory forms were found in Knossos, Crete. In Cyprus, Sparta and Mycenae game boxes, mirror grips and tablets carved with hunting or war scenes were discovered, possibly originating from Syria or the southern coast of Asia Minor. Finally, in the Classical Era, the combination of ivory and gold was represented by the enormous gold-and-ivory statues, sculptured by Phidias: Athena's in the Parthenon and Zeus' in Olympia, 5th century BCE.

---

[5] Pausanias 5, 12 6, Eustathius 1483.25, Plinius H.N. XXX, III, 23, Strabo 3.126, Herodotus I.50.

Figure 18.2 "Maternal love", a modern ivory art object of Japanese origin.

(a) (b)

Figure 18.3 Representation of the gold-and-ivory statues of (a) Athena of the Parthenon and (b) Zeus of Olympia.

18.1.4 Helmets

The Mycenaean helmets, at least in the early period, were made of boar tusks (Figure 18.4). The following account concerns the helmet of Odysseus:

Figure 18.4 Mycenaean armor with helmet Ileft) and a warrior armed with an eight-shaped shield and a helmet made of boar tusks (right), ivory bas-relief ca. 1400–1200 BCE from Delos, Delos Archaeological Museum.

> Μηριόνης δ’ ‘Οδυσῆϊ δίδου βιὸν ἠδὲ φαρέτρην
> καὶ ξίφος, ἀμφὶ δέ οἱ κυνέην κεφαλῆφιν ἔθηκε
> ρινοῦ ποιητήν· πολέσιν δ’ ἔντοσθεν ἱμᾶσιν
> ἐντέτατο στερεῶς· ἔκτοσθε δὲ λευκοὶ ὀδόντες
> ἀργιόδοντος ὑὸς θαμέες ἔχον ἔνθα καὶ ἔνθα
> εὖ καὶ ἐπισταμένως· μέσσῃ δ’ ἐνὶ πῖλος ἀρήρει.

> Meriones found a bow and quiver for Ulysses, and on his head he set a leathern helmet that was lined with a strong plaiting of leathern thongs, while on the outside it was thickly studded with boar's teeth, well and skillfully set into it; next the head there was an inner lining of felt.
>
> (Il. 10.260–265)

This description corresponds to a very balanced design: the natural product surrounding the helmet, boar tusks, has excellent impact strength, while the internal filling of felt absorbed impact preventing it to reach the skull.

Fully analogous is the construction of modern military helmets, manufactured from advanced composite. However, it appears that, at later times, composite laminates were used for the construction of helmets, just like the shields. Thus, their similarity with modern military helmets is even more striking (Figure 18.5).

Referring to boar tusks, the major part consists of *dentine*, a yellowish calcic substance, much denser and harder than bones. Dentine: (a) forms the main bulk of the tooth and is considered a vital tissue, just as the tooth pulp,[6] which provides nutrition, feeling and resistance to fracture, (b) provides the base for the much harder *enamel* and forms the root (or the roots) of the tooth, (c) is pierced by tubules, extending continuously from the pulp to the external surface, i.e., it is porous and, (c) is a tough material without preferable-direction fracture surfaces. The chemical composition of dentine b.w. is 70% inorganic, 20% organic and 10% water. The inorganic phase is oxyapatite and the organic is collagen. Its structure within the tooth is fairly complex.

The major part of the teeth of mammals is covered by *enamel*, which the hardest substance of the whole body. It consists mainly from apatite crystals, containing calcium and phosphoric salts. Enamel is harder at the points that the tooth bites.

18.2 The Panoply of Atreidēs

An account on the panoply of Agamemnon is given in Il. 11.15–46:

'Ατρεΐδης δ' ἐβόησεν ἰδὲ ζώννυσθαι ἄνωγεν
'Αργείους· ἐν δ' αὐτὸς ἐδύσετο νώροπα χαλκόν.
κνημῖδας μὲν πρῶτα περὶ κνήμησιν ἔθηκε
καλὰς ἀργυρέοισιν ἐπισφυρίοις ἀραρυίας·
δεύτερον αὖ θώρηκα περὶ στήθεσσιν ἔδυνε,
τόν ποτέ οἱ Κινύρης δῶκε ξεινήϊον εἶναι.
πεύθετο γὰρ Κύπρον δὲ μέγα κλέος οὕνεκ' 'Αχαιοὶ
ἐς Τροίην νήεσσιν ἀναπλεύσεσθαι ἔμελλον·
τοὔνεκά οἱ τὸν δῶκε χαριζόμενος βασιλῆϊ.
τοῦ δ' ἤτοι δέκα οἶμοι ἔσαν μέλανος κυάνοιο,
δώδεκα δὲ χρυσοῖο καὶ εἴκοσι κασσιτέροιο·
κυάνεοι δὲ δράκοντες ὀρωρέχατο προτὶ δειρὴν
τρεῖς ἑκάτερθ' ἴρισσιν ἐοικότες, ἅς τε Κρονίων

[6] Pulp is the live tissue within a tooth, placed in a special chamber and in the root channels. It contains, in high density, connecting tissue, nerves, lymph and blood vessels.

Figure 18.5 Mycenaean helmet manufactured from boar tusks (left), and a modern military helmet of Danish origin (right).

ἐν νέφεϊ στήριξε, τέρας μερόπων ἀνθρώπων.
ἀμφὶ δ’ ἄρ’ ὤμησιν βάλετο ξίφος· ἐν δὲ οἱ ἧλοι
χρύσειοι πάμφαινον, ἀτὰρ περὶ κουλεὸν ἦεν
ἀργύρεον χρυσέοισιν ἀορτήρεσσιν ἀρηρός.
ἂν δ’ ἕλετ’ ἀμφιβρότην πολυδαίδαλον ἀσπίδα θοῦριν
καλήν, ἣν πέρι μὲν κύκλοι δέκα χάλκεοι ἦσαν,
ἐν δέ οἱ ὀμφαλοὶ ἦσαν ἐείκοσι κασσιτέροιο
λευκοί, ἐν δὲ μέσοισιν ἔην μέλανος κυάνοιο.
τῇ δ’ ἐπὶ μὲν Γοργὼ βλοσυρῶπις ἐστεφάνωτο
δεινὸν δερκομένη, περὶ δὲ Δεῖμός τε Φόβος τε.
τῆς δ’ ἐξ ἀργύρεος τελαμὼν ἦν· αὐτὰρ ἐπ’ αὐτοῦ
κυάνεος ἐλέλικτο δράκων, κεφαλαὶ δέ οἱ ἦσαν
τρεῖς ἀμφιστρεφέες ἑνὸς αὐχένος ἐκπεφυυῖαι.
κρατὶ δ’ ἐπ’ ἀμφίφαλον κυνέην θέτο τετραφάληρον
ἵππουριν· δεινὸν δὲ λόφος καθύπερθεν ἔνευεν.
εἵλετο δ’ ἄλκιμα δοῦρε δύω κεκορυθμένα χαλκῷ
ὀξέα· τῆλε δὲ χαλκὸς ἀπ’ αὐτόφιν οὐρανὸν εἴσω
λάμπ’· ἐπὶ δ’ ἐγδούπησαν ‘Αθηναίη τε καὶ Ἥρη
τιμῶσαι βασιλῆα πολυχρύσοιο Μυκήνης.

The son of Atreus shouted aloud and bade the Argives gird themselves for battle while he put on his armour. First he girded his goodly greaves about his legs, making them fast with ankle clasps of silver; and about his chest he set the breastplate which Cinyras had once given him as a guest-gift. It had been noised abroad as far as Cyprus that the Achaeans were about to sail for

> Troy, and therefore he gave it to the king. It had[7] ten courses of dark cyanus, twelve of gold, and ten of tin. There were serpents of cyanus that reared themselves up towards the neck, three upon either side, like the rainbows which the son of Cronus[8] has set in heaven as a sign to mortal men. About his shoulders he threw his sword, studded with bosses of gold; and the scabbard was of silver with a chain of gold wherewith to hang it. He took moreover the richly-dight shield that covered his body when he was in battle- fair to see, with ten circles of bronze running all round see, wit it. On the body of the shield there were twenty bosses of white tin, with another of dark cyanus in the middle: this last was made to show a Gorgon's head, fiercc and grim, with Rout and Panic on either side. The band for the arm to go through was of silver, on which there was a writhing snake of cyanus with three heads that sprang from a single neck, and went in and out among one another. On his head Agamemnon set a helmet, with a peak before and behind, and four plumes of horse-hair that nodded menacingly above it; then he grasped two redoubtable bronze-shod spears, and the gleam of his armour shot from him as a flame into the firmament, while Juno and Minerva thundered in honour of the king of rich Mycene.

18.3 The Roman Shield

Polybius[9] in his History, Book 6, 23, provides a complete account of a Roman shield of his time. For reasons of comparison with Mycenaean technology, this description is given in full:

> The Roman *panoply* (i.e., armour) consists firstly of a shield (scutum), the *convex* surface of which measures two and a half feet in width and four feet in length, the thickness at the rim being a palm's breadth. It is made of two planks *glued* together with bull-glue,[10] the outer surface being then covered first with *canvas* and then with *calf-skin*. Its upper and lower rims are strengthened by an *iron* edging which protects it from descending blows and from injury when rested on the ground. It also has an *iron boss* (umbo) fixed to it which turns aside the most formidable blows of stones, *pikes*, and heavy missiles in general.

---

[7] Verse 24. This is the only detailed description of a breastplate in Homer (compare 23.560–562).

[8] Cronidēs, e.g., Zeus, son of Cronus.

[9] Polybius (ca. 200–118 BCE) Greek statesman and historian, who wrote of the development of Rome into a world power.

[10] Glue produced from bull hide.

Part 5
Further Issues

Chapter 19
The Trojan Horse

Doureios Hippos or the Trojan Horse was the huge wooden structure, built by the Achaeans on Odysseus' suggestion, hiding a number of fully armed select warriors inside. The Greeks, pretending to depart and give up Troy's siege, left it before the city walls as offering to the gods. The Trojans, discovering the "offering", had intense arguments: the suspicious ones maintained that the Greeks should not be trusted, but the pious insisted that the gods should receive what belonged to them. The latter prevailed and eventually it was decided to bring the horse into the city. To this end, they tore down part of the city-walls, to pass the huge structure through the opening. At night, the armed men emerged from their hideaway, overpowered the guards and opened the gates. The lurking Greeks entered the city and conquered it. The events are related in the Odyssey, in the court of Alcinous, king of the Phaeacians, where Demodocus, the bard, was singing of the Trojan War in the presence of Odysseus. The specific passage follows:

> αὐτὰρ ἐπεὶ πόσιος καὶ ἐδητύος ἐξ ἔρον ἕντο,
> δὴ τότε Δημόδοκον προσέφη πολύμητις ‘Οδυσσεύς·
> “Δημόδοκ’, ἔξοχα δὴ σὲ βροτῶν αἰνίζομ’ ἁπάντων
> ἢ σὲ γε Μούσ’ ἐδίδαξε, Διὸς πάϊς, ἢ σὲ γ’ ’Απόλλων·
> λίην γὰρ κατὰ κόσμον ‘Αχαιῶν οἶτον ἀείδεις,
> ὅσσ’ ἔρξαν τ’ ἔπαθον τὲ καὶ ὅσσ’ ἐμόγησαν ’Αχαιοί,
> ὥς τέ που ἢ αὐτὸς παρεὼν ἢ ἄλλου ακούσας.
> ἀλλ’ ἄγε δὴ μετάβηθι καὶ ἵππου κόσμον ἄεισον
> δουρατέου, τον ’Επειὸς ἐποίησεν σὺν ‘Αθήνῃ,
> ὃν ποτ’ ἐς ἀκρόπολιν ἤγαγε δῖος ‘Οδυσσεὺς
> ἀνδρῶν εμπλήσας, οἵ ῥ’ ’Ίλιον ἐξαλάπαξαν.
> αἴ κεν δὴ μοι ταῦτα κατὰ μοῖραν καταλέξῃς,
> αὐτίκ’ ἐγὼ πᾶσιν μυθήσομαι ἀνθρώποισιν
> ὡς ἄρα τοι πρόφρων θεὸς ὤπασε θέσπιν ἀοιδήν.”
> ὥς φάθ’, ὁ δ’ ὁρμηθεὶς θεοῦ ἤρχετο, φαῖνε δ’ ἀοιδὴν

S.A. Paipetis, *The Unknown Technology in Homer*, History of Mechanism and Machine Science 9, 169
DOI 10.1007/978-90-481-2514-2_19, © Springer Science + Business Media B.V. 2010

ἔνθεν ἑλὼν ὡς οἱ μὲν ἐϋσσέλμων ἐπὶ νηῶν
βάντες ἀπέπλειον, πῦρ ἐν κλισίῃσι βαλόντες,
Ἀργεῖοι, τοὶ δ' ἤδη ἀγακλυτὸν ἀμφ' Ὀδυσῆα
ἥατ' ἐνὶ Τρώων ἀγορῇ κεκαλυμμένοι ἵππῳ·
αὐτοὶ γάρ μιν Τρῶες ἐς ἀκρόπολιν ἐρύσαντο.
ὣς ὁ μὲν ἑστήκει, τοὶ δ' ἄκριτα πόλλ' ἀγόρευον
ἥμενοι ἀμφ' αὐτόν· τρίχα δέ σφισιν ἥνδανε βουλή,
ἠὲ διαπλῆξαι κοῖλον δόρυ νηλέϊ χαλκῷ,
ἢ κατὰ πετράων βαλέειν ἐρύσαντας ἐπ' ἄκρης,
ἢ ἐάαν μέγ' ἄγαλμα θεῶν θελκτήριον εἶναι,
τῇ περ δὴ καὶ ἔπειτα τελευτήσεσθαι ἔμελλεν·
αἶσα γὰρ ἦν ἀπολέσθαι, ἐπὴν πόλις ἀμφικαλύψῃ
δουράτεον μέγαν ἵππον, ὅθ' ἥατο πάντες ἄριστοι
Ἀργείων Τρώεσσι φόνον καὶ κῆρα φέροντες,
ἤειδεν δ' ὡς ἄστυ διέπραθον υἷες Ἀχαιῶν
ἱππόθεν ἐκχύμενοι, κοῖλον λόχον ἐκπρολιπόντες.
ἄλλον δ' ἄλλῃ ἄειδε πόλιν κεραϊζέμεν αἰπήν,
αὐτὰρ Ὀδυσσῆα προτὶ δώματα Δηϊφόβοιο
βήμεναι, ἠΰτ' Ἄρηα, σὺν ἀντιθέῳ Μενελάῳ.
κεῖθι δὴ αἰνότατον πόλεμον φάτο τολμήσαντα
νικῆσαι καὶ ἔπειτα διὰ μεγάθυμον Ἀθήνην.

They then laid their hands on the good things that were before them, and as soon as they had had to eat and drink, Ulysses said to Demodocus, "Demodocus, there is no one in the world whom I admire more than I do you. You must have studied under the Muse, Jove's daughter, and under Apollo, so accurately do you sing the return of the Achaeans with all their sufferings and adventures. If you were not there yourself, you must have heard it all from some one who was. Now, however, change your song and tell us of the wooden horse which Epeus made with the assistance of Minerva, and which Ulysses got by stratagem into the fort of Troy after freighting it with the men who afterwards sacked the city. If you will sing this tale aright I will tell all the world how magnificently heaven has endowed you." The bard inspired of heaven took up the story at the point where some of the Argives set fire to their tents and sailed away while others, hidden within the horse, were waiting with Ulysses in the Trojan place of assembly. For the Trojans themselves had drawn the horse into their fortress, and it stood there while they sat in council round it, and were in three minds as to what they should do. Some were for breaking it up then and there; others would have it dragged to the top of the rock on which the fortress stood, and then thrown down the precipice; while yet others were for letting it remain as an offering and propitiation for the gods. And this was how they settled it in the end, for the city was doomed when it took in that horse, within which were all the bravest of the Argives waiting to bring death and destruction on the Trojans. Anon he sang how the sons of the Achaeans issued from the horse, and sacked the

Figure 19.1 The Mykonos vase, early Archaic Period (ca. 670 BCE), the first artifact found to depict the Trojan Horse.

> town, breaking out from their ambuscade. He sang how they over ran the city hither and thither and ravaged it, and how Ulysses went raging like Mars along with Menelaus to the house of Deiphobus. It was there that the fight raged most furiously, nevertheless by Minerva's help he was victorious.
>
> (Od. 8.485–520)

The first depictionn of the Trojan Horse was found on the Mykonos Vase, discovered in 1961 in the Greek island of Mykonos, in the Aegean, by a local resident (Figure 19.1). The Trojan Horse and its story have been a great source of inspiration to artists through the centuries (Figures 19.2 and 19.3).

Assuming that the Trojan Horse was a real structure and not just a poetic concept, its specifications and technical requirements need to be examined:

(a) Although no number of the men in the horse is specified (references tell of 30 to 3000!), it is estimated that, for the operation to have any chance of success at all, at least 100 heavily armed warriors or commandos should be hidden in the Horse, to take the Trojan guards by surprise.
(b) According to the narrative, the warriors should remain in the horse for at least five days, to account for the time the horse had to stay out of the city, the time of discussions among the Trojans, the demolition of part of the walls and finally the transportation of the structure into the city, a difficult task for such a heavy object.

(a)

(b)

Figure 19.2 Giovanni Domenico Tiepolo (1727–1804): (a) "The construction of the Trojan Horse", and (b) "The introduction of the Wooden Horse into Troy", ca. 1760, National Gallery, London. The paintings belong to a series of sketches, each one of which illustrates a scene from Virgil's Aeneid (Book 2) respectively: "The Danaans constructed a wooden horse, they put armed men into it and left it in front of the enemy city of Troy", and "The Trojans, who adored horses, bring the gift of the Danaans into the city. Later, the Danaan warriors emerge and start destroying Troy" (reproduced by permission).

Figure 19.3 Two paintings entitled "The Trojan Horse" by Raoul Lefevre (1464, left) and W. Friedrich (ca. 1894, right).

(c) Each warrior along with his armor and equipment should weigh at least 100 kg, plus food and water for five days, the total net weight should be of the order of 10–12 tons.

(d) The structure itself should have ventilation openings and a waterproof drainage system, sealed against odors that might arouse suspicions of its contents.

(e) The inner arrangement should resemble that of a modern long distance passenger coach or airplane cabin, consisting of seat rows with corridors at either side, to enable distribution of food and water, access to the toilet or to some kind of physical activity to keep the confined men in good shape, ready to rush out and take immediate action. In any case, masses should be properly arranged, to obtain a well-balanced and stable structure, not risking the structures turning over during transportation.

(f) The seat rows should have the appropriate length, so that the whole structure remain, even roughly, geometrically similar to a horse and its analogies (Figure 19.4), which imposes further restrictions to the arrangement of the persons, as, for example, to have two-seat rows, e.g., one on top

Figure 19.4 Picture of a horse showing the analogies of its body. Species Equus caballus, sole member of Equidae family. As a domestic animal, may reach up to 2 m in height.

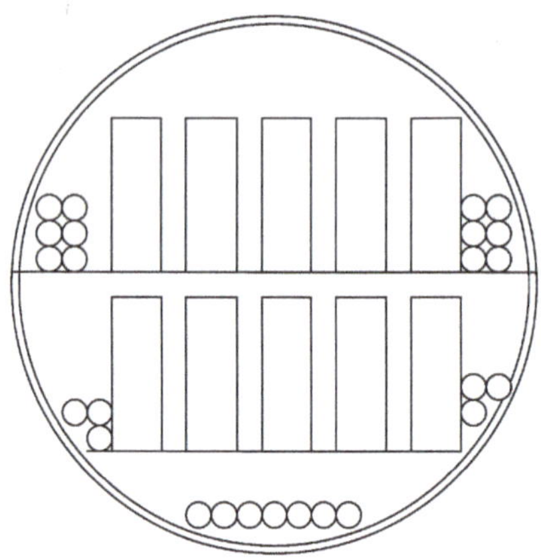

Figure 19.5 Cross section of the body of the Trojan Horse considered as a cylinder and a potential arrangement of persons (rectangles) and equipment (circles). Based on the physical dimensions of the human body, etc., its diameter is estimated to around 2.5 m.

of the other (Figure 19.5). In other words, the inner arrangement of seats would be like that of a modern two-level coach, but with increased facilities of storing arms and equipment and raised above the ground by the height of the horse legs. Dimensions are calculated in the sequence.

(g) As derived from Figure 19.5, based on a 2.5 m structure diameter on the analogies of Figure 19.4, the general dimensions are calculated as follows:

- Total height of the structure: 6.25 m (as a two-store apartment building).
- Length of cylindrical (effective) part: 6.25 m.
- Length from the neck base to the end of the head: 2.25 m.
- Total length of the structure (from head to tail): 8.75 m.
- Height of legs: 3.75 m.

Based on these dimensions, an elementary structural analysis and evaluation of the total weight are performed in the sequence.

19.1 Wood as Structural Material

The use of wood as a structural material is governed by its mechanical properties, mainly by its strength against loads tending to change its shape and size. The load-carrying capacity of wooden structures depends both on the intensity and mode of application of these forces and on the properties of wood, out of which density and humidity content are very important. It must be emphasized that wood is strongly anisotropic, i.e., it exhibits completely different strength in the direction of its fibres than normally to it. Therefore, its mechanical properties must have been determined in both directions. These particular properties are strength in tension and in compression, shear strength, static bending strength, impact strength or toughness etc. The strength of wood increases with density and decreases with increasing humidity as well as with increasing temperature. Long-time loading reduces its strength, while of critical importance are all kinds of faults, depending on their kind, size and position in relation with the loads applied.

According to the above, the Trojan Horse was not a simple wooden structure. It had very demanding short-term requirements, although it was not designed for long operational life. Wood should have high strength, to minimize cross sections and weight of loaded structural elements, i.e., dry wood should be employed, which exhibits maximum strength in all cases. In this specific case, wood for shipbuilding would be ideal, such as fir tree wood, which, being available in great lengths, was suitable for large structural elements. Moreover, the Greeks would certainly carry in their ships sufficient inventories of such wood, to repair battle damages. No wood from local trees was required.

Fir is repeatedly mentioned in Homer, praising its good properties. It is named ούρανομήκης (= *sky-high*) and, in fact, a famous fir of Mount Ida is mentioned, indeed reaching the sky:

> ἔνθ' Ὕπνος μὲν ἔμεινε πάρος Διὸς ὄσσε ἰδέσθαι
> εἰς ἐλάτην ἀναβὰς περιμήκετον, ἣ τότ' ἐν Ἴδῃ
> μακροτάτη πεφυυῖα δί' ἠέρος αἰθέρ' ἵκανεν.
>
> Here Sleep halted, and ere Jove caught sight of him he climbed a lofty pinetree – the tallest that reared its head towards heaven on all Ida.
>
> (Il. 14. 286–288)

Also, for its use in shipbuilding:

> ... ἦρχε δ' ὁδοῖο
> νήσου ἐπ' ἐσχατιῆς, ὅθι δένδρεα μακρὰ πεφύκει,
> κλήθρη τ' αἴγειρός τ', ἐλάτη τ' ἦν οὐρανομήκης,
> αὖα πάλαι περίκηλα, τὰ οἱ πλώοιεν ἐλαφρῶς.
>
> ... and then led the way to the far end of the island where the largest trees grew – alder, poplar and fir, that reached the sky – very dry and well seasoned, so as to sail light for him in the water.
>
> (Od. 5.237–240)

The trunk of fir tree was suitable for ship masts:

> ἱστὸν δ' εἰλάτινον κοίλης ἔντοσθε μεσόδμης
>
> They set the fir mast in its socket in the cross plank, raised it
>
> (Od. 2.424)

Pitys, another tree useful in shipbuilding, mentioned by Homer, is a kind of pine tree, which Theophrastus calls *pine of Ida*, since it grows in the area of Mount Ida (*Pinus Laricio*). There are the following references:

> ἤριπε δ' ὡς ὅτε τὶς δρῦς ἤριπεν ἢ ἀχερωῒς
> ἠὲ πίτυς βλωθρή, τὴν τ' οὔρεσιν τέκτονες ἄνδρες
> ἐξέταμον πελέκεσσι νεήκεσι νήϊον εἶναι·
>
> He fell as an oak, or poplar, or pine which shipwrights have felled for ship's timber upon the mountains with whetted axes
>
> (Il. 13.389–391)

(see also Il. 16.482–484). Also

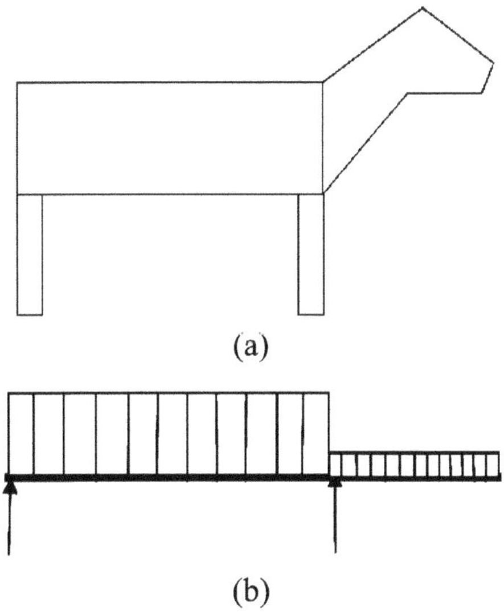

Figure 19.6 (a) Two-dimensional model of the body of Trojan Horse, (b) considered as a simple beam.

μακρῇσίν τε πίτυσσιν ἰδὲ δρυσὶν ὑψικόμοισιν.

with tall pine trees and tall-trunk oaks

(Od. 10.186)

For this tree, the same remarks hold.

19.1.1 An Elementary Structural Analysis

For an elementary structural analysis, the simplified two-dimensional model of Figure 19.6 was devised, by which the horse body was considered fully cylindrical, with the cross section of Figure 19.6a and uniformly loaded. The horse head should be hollow, to store supplies. Its weight was taken equal to 1 ton or 10% of the total weight and is uniformly distributed along its length.

The whole structure was modelled as a simple beam, Figure 19.6b. Elementary calculations, performed on the basis of strength values for dry fir tree wood, led to a total gross weight of the order of 10–12 tons. Not a high-tech structure, which was well within the technical capabilities of the time, which, however, are proved to be extraordinarily advanced, if not, for that matter, admirable.

Chapter 20
Mycenaean Building

Since the middle of the 13th century BCE, colossal defensive walls were protecting the various palaces of the Greek mainland, exhibiting tremendous skill in using huge rocks in the design of complex gates and protection systems of underground water-supply networks. In Tiryns, the walls are possessing elegant recesses, while in Mycenae the famous Lions' Gate is decorated with a sculpture of two lions on either sides of a column. Such fortifications probably indicate frictions and conflicts between city-states, characterizing ancient Greece or measures against enemies so far unknown.

Although such fortifications should be very costly, palaces of the 13th century BCE were built with increasing size and architectural quality, while

Figure 20.1 Tiryns castle.

S.A. Paipetis, *The Unknown Technology in Homer*, History of Mechanism and Machine Science 9, DOI 10.1007/978-90-481-2514-2_20, © Springer Science + Business Media B.V. 2010

walls and floors appeared to be reconstructed. Burial offerings are highly valuable, indicating wealth. Assuming that both vaulted tombs in Mycenae, i.e., Atreus' Treasury and Clytemnestra's Tomb were built during this latter period, then respective dynasties must have been possessing unlimited resources.

Mycenae was a very important centre in 2nd millennium BCE Greece. Archaeological research has revealed numerous settlements of that era and a huge trade and cultural exchange network, characteristic of the Mycenaean world and constituted the Greek forces that took part in the Trojan War.

Schliemann's next target, after Troy, was Mycenae, the kingdom of Agamemnon, the commander-in-chief of the Greek forces of the Trojan War. He did not have to try equally hard as in Troy. The place was well known and the ancient ruins were by a large part visible.

Schliemann's first important findings were huge circular arrangements of graves (Figure 20.2), containing numerous burial offerings, and so on, of great value, as shown in Figures 20.3 to 20.5.

The enormous size of the city of Mycenae is shown in Figure 20.6. A circular arrangement of tombs lies at the lower part, where the death masks and other golden items were found. The royal palace was in the upper part. The acropolis is surrounded by the massive *Cyclopean Walls*, which, combined with the strategic position of the city, rendered it practically inconquerable.

The Lion's Gate was the entrance to the castle and the civil centre of Mycenae, presented in Figures 20.7 and 20.8. Note the huge size of the rocks and the narrow opening, confirming that the engineers of the time were unaware of the principles of arch design, which would allow for wider openings.

20.1 The Treasury of Atreus

The so-called Atreus' Treasury lies near the ruins of Mycenae, a vaulted structure shown in Figures 20.9 and 20.10. Initially, it was thought to be a treasury belonging to Atreus, father of Agamemnon, however, most probably it was a royal grave. A description of the monument is given at the site of the Ministry of Culture of Greece.[1] The importance of Atreus' vaulted tomb is fully described by Mylonas,[2] as follows:

[1] http://www.culture.gr/2/21/211/21104n/e211dn01.html

[2] Mylonas G.E., *Mykenae – Rich in Gold*, Ekdotikē Athinon SA, Athens 1983.

Figure 20.2 Mycenaean graveyards.

Figure 20.3 Death masks of Mycenaean graves made of pure gold, Archaeological Museum, Athens.

Figure 20.4 The famous Mycenaean death mask, which, according to Schliemann, belonged to Agamemnon himself, Archaeological Museum, Athens.

Figure 20.5 From the Mycenaean treasures: Golden covers for the body of a dead infant (left) and a golden cup, which, according to Schliemann, belonged to Nestor (right).

A few years ago, I wrote of my experience with the Monument (e.g., Atreus' vaulted tomb), as I was studying it in daylight or with an electric searchlight. Perhaps, I am entitled to repeat here some of my findings of the first years, which may correspond to thoughts that will flood the mind of a modern visitor during his pilgrimage in the grave, in which leaders of the Heroic Era of Greece have been buried.

A simple glance at the Treasury-Tomb of Atreus shall be enough to prove that this is one of the most impressive monuments of the Mycenaean World.

Figure 20.6 Air photograph of the ruins of the city of Mycenae (left) and reconstruction under the same perspective (right).

Figure 20.7 Painting of 1810, depicting the Lions' Gate half-buried, i.e., before excavation.

The Cyclopean Walls, the Lions' Gate, the Underground Cistern attract the admiration of the eye, but what excites imagination and the tendency towards the imaginary within the human mind is the proud erection of Atreus' Treasury Tomb. The design and dexterity of construction, its proportions and the imposing tomb, the care with which the rings of the vault were put to place and polished, the precautions taken against permeance of rain water, the obvious skill by which huge cubed rocks were hewn to form a round structure and left in place, all witness the high degree of perfection reached by the architects and the stone builders of Mycenae during the 14th and 13th centuries BCE.

Figure 20.8 The Lions' Gate as it is today.

> We wish that it were possible for us to depict the ritual of burial, as it took place within the magnificent vault, to reconstruct its interior, as it was when the body of the last king was placed in the ground or in the adjacent chamber, and when the solid stone wall was built across the opening! But the contents of the tomb should have gone long before the time of Pausanias, and what has remained are the various depictions that fantasy can give us, as excited by silence and by the impressive architectural remnants, a work of great men.

In ancient times, vaulted subterranean structures, were not necessarily hewn in the ground or in rock. On many occasions they were *a posteriori* covered with earth, forming a *tymvos*, i.e., a little hillock. Most probably the earth played a practical role: In the absence of efficient hoisting or lifting machines, capable of raising the Cyclopean rocks at the desired height, this role was taken over by a continually added amount of earth, operating as a ramp, on which rocks could slide. When the building was completed and depending on its operational use, earth could be removed or even remain in place.

Figure 20.9 Entrance to Atreus' Treasury and a sketch of its interior.

The specific vault, as opposed to usual structures, has a side corridor sculptured in the rock, which, according to local tradition, was covering Agamemnon's golden chariot.

The main vault has a 14.2 m diameter at the ground and 13.2 m height, e.g., its diameter is almost equal to its height. It consists of 33 successive rings of conglomerate stones, perfectly interconnected without any plaster. The solid floor is covered with compacted hard whitish ground.

The vault's façade is maintained in very good condition, although its decorations have been stolen. Some of its fragments exist in the British Museum, as parts of the marbles removed by Elgin, also in the Athens Archaeological Museum and in Munich, Karlsruhe and Berlin Museums. A reconstruction of the façade is shown in Figure 20.10.

In Figure 20.11 a detail is shown of the construction of a door opening at the northern side of the vault, very characteristic of the Cyclopean technique used. This structure seems to possess extraordinary properties of strength and stability. The fact was proved by a study of its static and dynamic behaviour performed by a research team of Civil Engineering Department, University of Patras[3] by using modern numerical methods.

---

[3] P. Askouni, M. Sfakianakis and D. Beskos, Static and dynamic analysis of Atreus' vaulted tomb in Mycenae, MSc Dissertation, Civil Engineering Department, University of Patras, 2004. See also P.K. Askouni, X.A. Angelopoulou, M.G Sfakianakis and D.E. Beskos, Static and dynamic analysis of Atreus' Tomb in Mycenae, in *Science and Technology in Homeric Epics*, S.A. Paipetis (Ed.), Springer, 2008. The author wishes to thank the authors of this most interesting work for making their results available for publication in the present book.

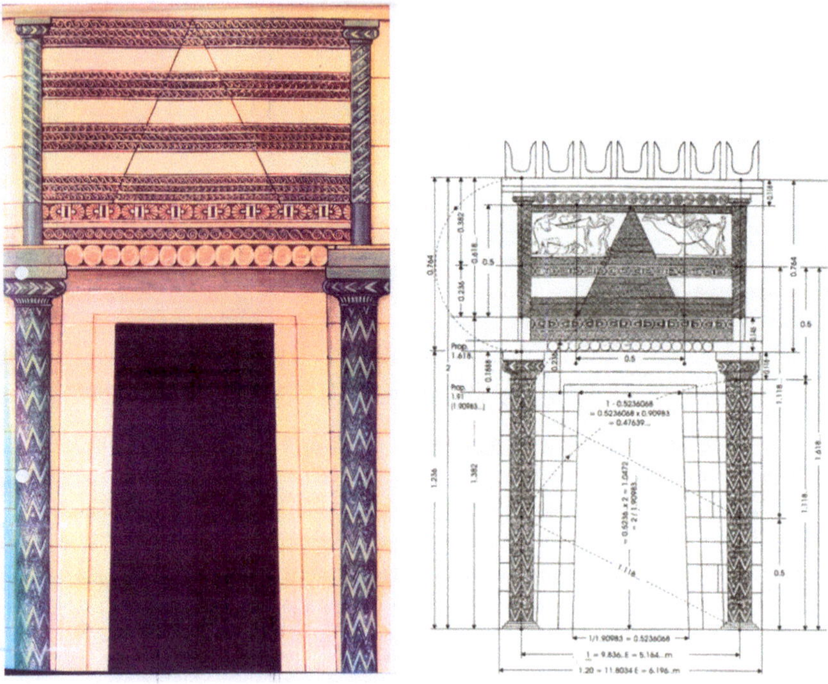

(a) (b)

Figure 20.10 (a) Reconstruction of the façade of the Treasury[4] as erected in 1250 BCE, and (b) a metrological analysis of its dimensions[5] (the decorations have been added schematically).

In the study, the anelastic behaviour of wall material is dealt with, based on damage theory, while for the analysis, a general static and dynamic analysis code was applied, using three-dimensional finite elements (FE).[6] The FE mesh for the analysis of the vaulted tomb along with the coordinate axes for the main entrance view and for plan-view appears in Figure 20.12.

The analysis was based on realistic values for the mechanical properties both of the structural materials and the overlying earth layers. In fact, both cases were examined, i.e., with and without filling.

The analysis was aiming at the determination of mechanical stresses at various points of the structure and the presentation of stress contours in the

[4] Marinatos, S., *Small Research in Mycenae*, AE 1953-54, 9-24 (in Greek).

[5] Kamm, Walther, Treasury of Atreus: The construction of the façade, in *Extraordinary Machines and Structures in Antiquity*, S.A. Paipetis (Ed.), Peri Technon Publ., Patras, 2002.

[6] M. Sfakianakis, MINOS 3-D: Programmer's and User's Guide.

Figure 20.11 Detail from the main entrance of the vault, showing the "Cyclopean technique" and the size of the rocks used.

colour diagrams that follow (Figure 20.13), referred to the above Cartesian coordinates.

The characteristics of the said diagrams are presented in Table 20.1, where all contour values are numerically maximum tensional or maximum compressional at all time steps.

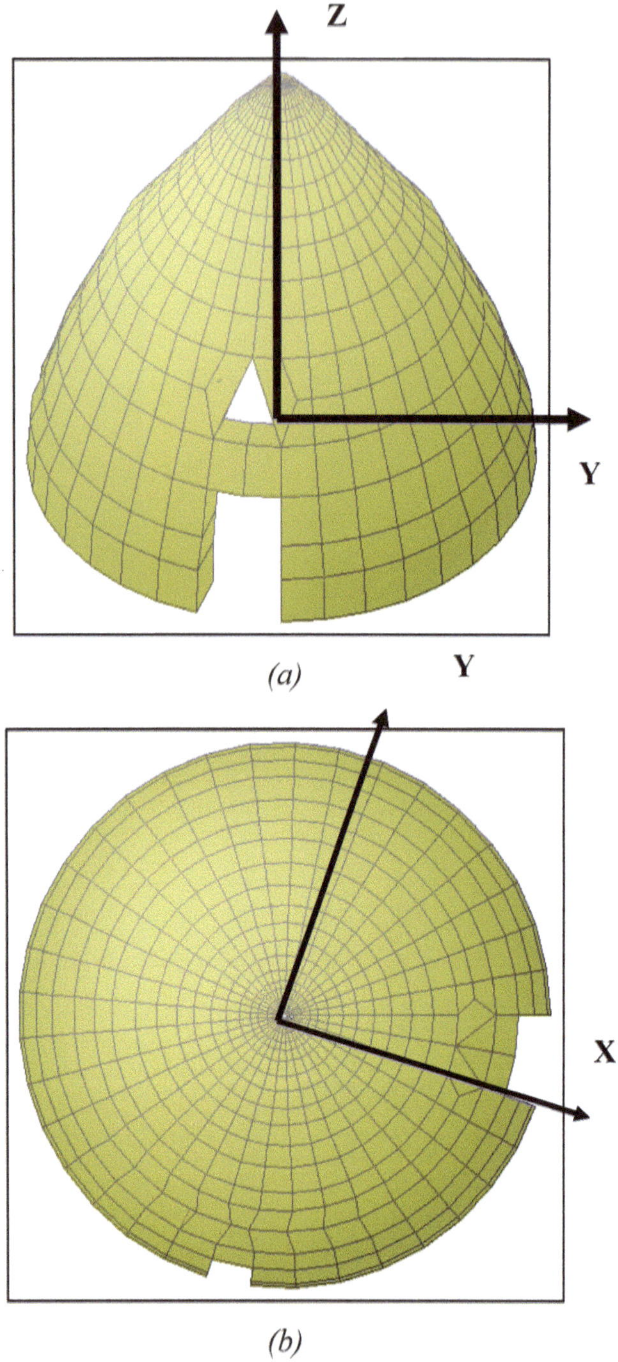

Figure 20.12 Finite-element mesh for the analysis of the vaulted tomb. (a) Face of main entrance and (b) plan view (picture from MINOS-Plot).

Table 20.1

s/n face pairs	Stresses σ	Filling scale factor	El-Centro[7] factor a_m	Mass damping
No. 1	$\sigma x(+)$	NO	1.00	0.00
No. 2	$\sigma x(-)$	NO	1.00	0.00
No. 3	$\sigma z(+)$	NO	1.00	0.00
No. 4	$\sigma z(-)$	NO	1.00	0.00
No. 5	$\sigma x(+)$	YES	1.00	0.00
No. 6	$\sigma x(-)$	YES	1.00	0.00
No. 7	$\sigma z(+)$	YES	1.00	0.00
No. 8	$\sigma z(-)$	YES	1.00	0.00

From the diagrams in Figure 20.13 it may be concluded that, even in the most unfavourable case, the stresses developing in the structure do not exceed 15% of the tensile strength and about 11% of the compressive strength. Therefore, no damage may occur at any point of the structure, especially, in view of the fact that all stresses developing remain elastic till the end, whether the loading is static or dynamic (high magnitude eartquake).

It is noted that the above analysis concerns the monument as it is today, while the techniques and means used for the erection of this amazing structure, which remains substantially untouched for over 3000 years, are still unknown, remaining, however, a wide field of further research.

[7] The town of El Centro was established in 1907 and is the seat of Imperial County in southwestern California, USA, 193 km east of San Diego. It is the largest US settlement situated lower than sea-level, e.g., at a −16 m negative altitude. In May 1940, an earthquake of magnitude 7.1 of the Richter scale caused damage to varying degrees to more than 80% of the buildings of Imperial County, while, in certain areas, the destruction of the buildings was almost complete. The tremor was caused by the 40-mile long surface fault on the Imperial Fault, which is part of the San Andreas system in Southern California. Material losses exceeded $6M. The characteristics of the event were recorded in detail, and, ever since, they are used as standards to check dynamic behaviour of structures.

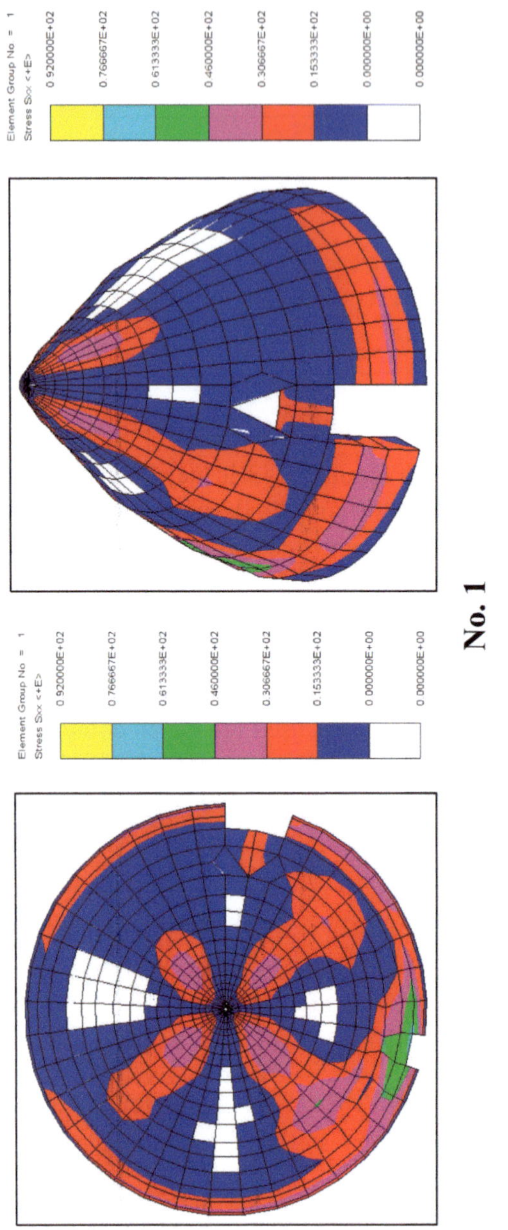

Figure 20.13 Stress contours at the various points of the structure based on El Centro earthquake.

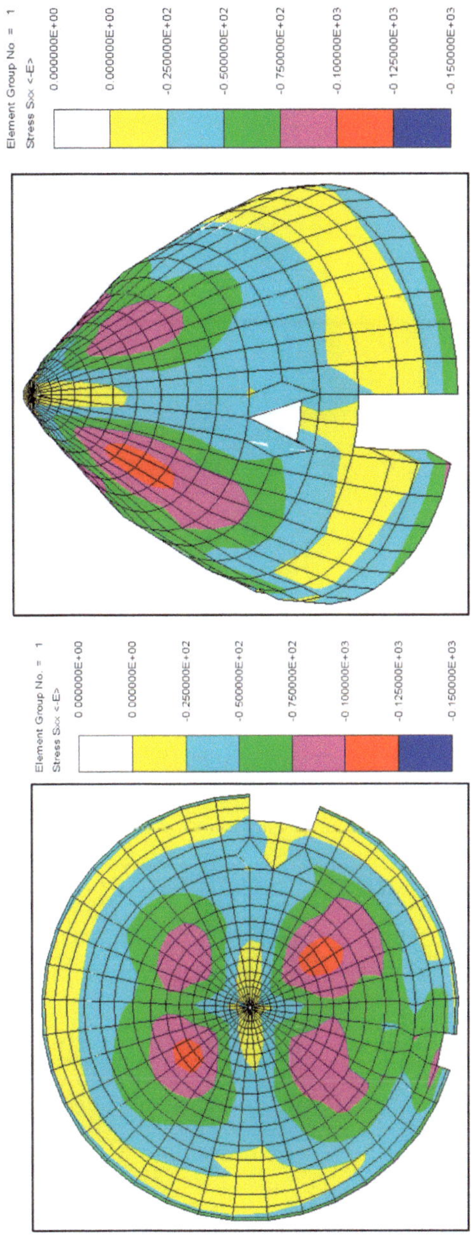

No. 2

Figure 20.13 Continued

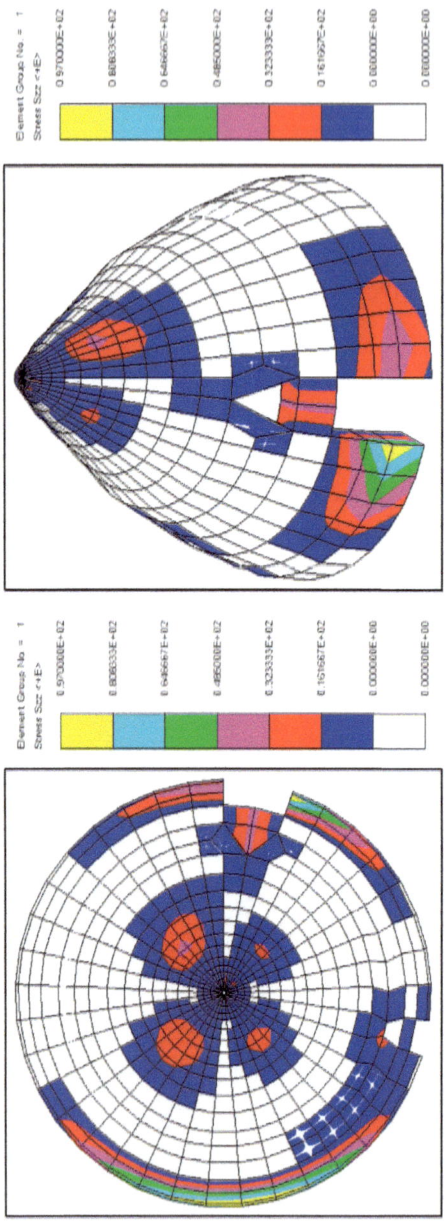

No. 3

Figure 20.13 Continued

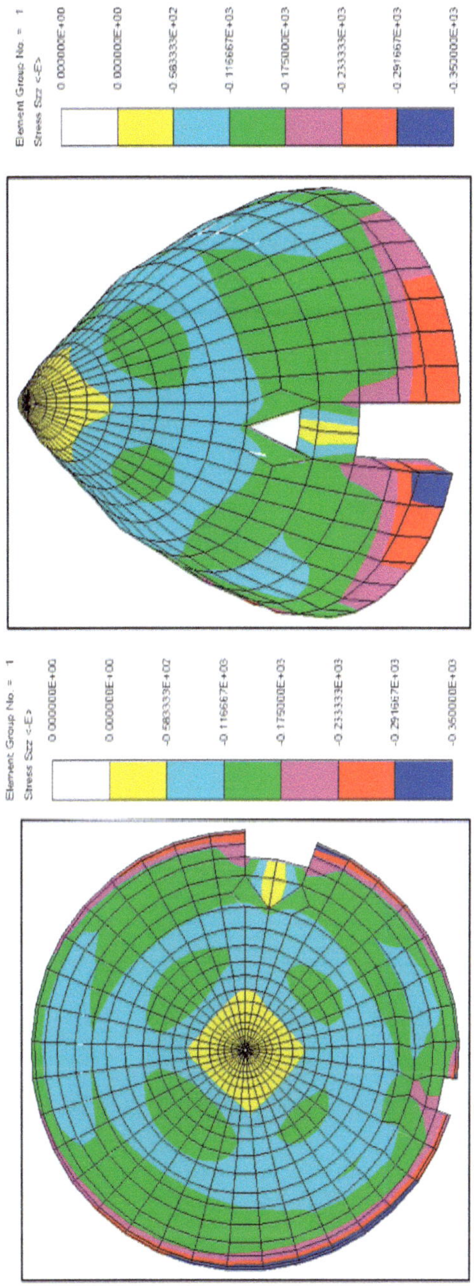

No. 4

Figure 20.13 Continued

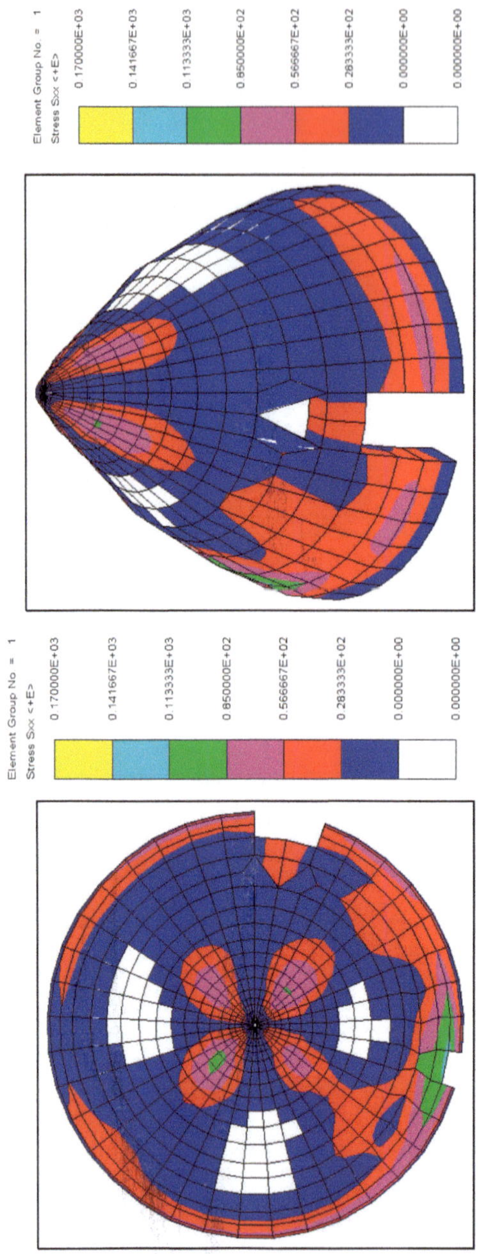

No. 5

Figure 20.13 Continued

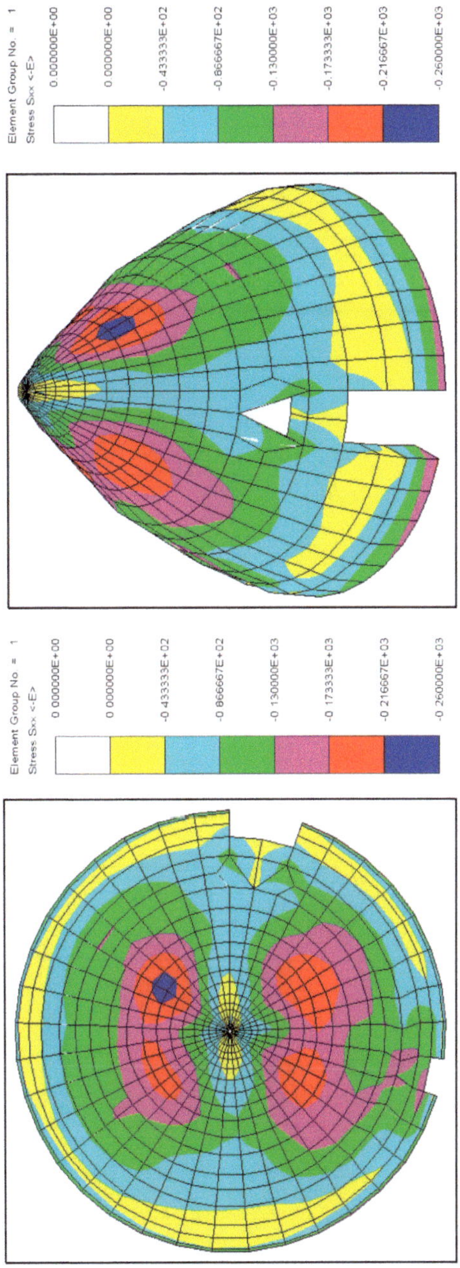

No. 6

Figure 20.13 Continued

Chapter 21
The Miraculous Homeric Meter

The particular structure of the poetic word used in the Homeric Epics, has led to a phenomenally paradoxical statement: Homer can help our heart. An equally paradoxical response of several scientists in Europe and the USA is "by *reciting the Odyssey*!" This is not a witticism. Recitation of the Homeric Epics with their proper metre causes synchronization of heart and respiration rates, a beneffficient effect, similar to the effect of Christian prayer using rosaries or Hindu and Buddhist yoga meditation using various mantras. In other words, the Homeric Epics can be used in the same way that the sacred scriptures of various religions and esoteric traditions are used for meditation.

21.1 Meditation

By this term one means either individual worship action or a mental exercise, consisting of many different techniques for concentration, contemplation and subtraction, supposedly leading to a higher spiritual realization or bodily relaxation.

Exercising meditation is a most ancient and universal practice in many different ambiences. It may serve relaxing (ἡσυχαστικούς) purposes, as in the case of hermits, as well as a method of rehabilitation and enrichment of every-day life, which is the case with numerous religious and secular bodies and individuals. Also, in the form of concentration, in view of an extreme effort or trial, such as before a tough game, theater performance or examination. In any case, according to recent medical and psychological studies, meditation techniques are substantially helping trained individuals to control heart and respiration rates and, to varying degrees, to control disturbing

S.A. Paipetis, *The Unknown Technology in Homer*, History of Mechanism and Machine Science 9,
DOI 10.1007/978-90-481-2514-2_21,

symptoms of syndromes, such as migraine headaches, high blood pressure, haemophilia, etc.[1]

Many of the great religions have developed their own meditation schools, such as Hindu yoga, Tibetan and Japanese Zen, etc. Respective techniques consist of *spoken* (legomena), *shown* (deiknymena) and *acted upon* (drōmena) elements.

As *legomena*, besides chanting and music, one is using special syllables, words or full phrases, which oriental religions call *mantra* (in Sanskrit "mental tool"), islam *dhikr*, while in Christianity the role is taken up by specific phrases, such as "Kyrie eleison" or "Ave Maria", etc., repeatedly uttered for long time. At the first stages of meditation, a novice learns how to control, in fact, how to *decrease* his/her brain activity and concentrate on respiration rate or use as mantra typical words, creating no associations that might induce thought trains. On the contrary, at advanced contemplation levels, meditation is supposed to lead to direct communication with the divine, practically developing into prayer, with full repetitive phrases. The supreme mantra in Hindu meditation is AUM, believed to be the creative sound of the Universe, whose three letters correspond to birth, conservation and destruction, as expressed by the Hindu divine triad (Brahma, Vishnu, Siva). The respective word in Tibetan Buddhism is OM, which is part of the equally important mantra *Om Mani Padme Hum* (Figure 21.1).

With *deiknymena*, meditation is focusing attention on pictures, depicting, for example, a flower or a mountain. In many traditions they assume typical forms, as in Tibetan Tantric Buddhism,[2] where a *mandala* (in Sanskr. *circle)* is considered as a concentration point of universal forces, which a human may contact through meditation (Figure 21.2).

For meditation, many traditions are using objects or mechanical devices, such as rosaries and the prayer wheel. Finally, *drōmena* are various motions or gestures, walking etc. synchronized with the recitation of a mantra.

---

[1] The author, out of personal experience, can confirm that, through meditation, recurrent migraine can be alleviated or even disappear, and also that a strong back-ache was cured instantly, apparently after successful relaxing of the respective nerve.

[2] Tantras (in Sanskrit "loom"), any of the numerous scriptures dealing with esoteric practices of certain Hindu, Buddhism or Jaina cults. Buddhist Tantras date back to the 7th century or even earlier. Tathāgataguhyaka is an early and extreme work. Tantras have been translated into Tibetan and Chinese from 9th century on. Only a few texts in these languages are extant, while the Sanskrit originals have been lost. An important text among Buddhist Tantras is Kālacakra-tantra.

(a) (b)

Figure 21.1 (a) The Hindu mantra AUM[3] (Sanskr.), (b) the Tibetan OM.

Decreased brain activity can be obtained by exclusion of thought trains and focusing on a mantra and, as mentioned, may have benevolent effects on the human body. The weight of human brain is only 2% of that of the body, however, it absorbs 16% of blood and 20% of the oxygen it carries (Figure 21.3).

It was proved by EEG that the brain remains active during sleep. Decreased thinking activity, achieved through systematic meditation, causes decreased brain energy consumption, e.g., less blood circulation, lower metabolism rate and general relaxation of the body. Similar is the change of brain waves, whereby *Alpha waves* (frequency 10 cycles per second) prevail with respective decrease of the irregular, noise-like *beta* waves. Very low-frequency *theta* waves appear with deep meditation. Finally, results similar to those of meditation can be obtained by *Biofeedback*.[4]

[3] The mantra in Sanskrit appears on a PC screen if one inputs reverse slash \ and convert it into Microsoft *Wingdings* font.

[4] Biofeedback: Information instantly supplied to a person in relation to his/her own physiological parameters, e.g., blood pressure, heart rate, body temperature, brain wave of muscle tension. This information, in the form of an electronic signal, is returned to the person through a measuring element or a light or sound monitoring. In this way the autonomous neural system is "bridged" with the thinking process, so that a trained individual can control the involuntary body functions, for example, to decrease the symptoms of an ailing, such as pain or muscle tension, but also migraine headaches, colitis, blood hypertension, nervous ticks, as well as frequency and intensity of epileptic fits. Through feedback of brain waves, the brain functions are enhanced. In particular, it generates all tranquillizing and holistic effects of meditation, while training with theta waves leads to improved attention focusing and control of mental hindrances and stress.

Figure 21.2 Tibetan Buddhist mandala used for meditation, representing the Universe and also the Temple or the City. In the external triangles are residing deities, symbolizing subdivisions of the energy essence of the Great Godess.[5]

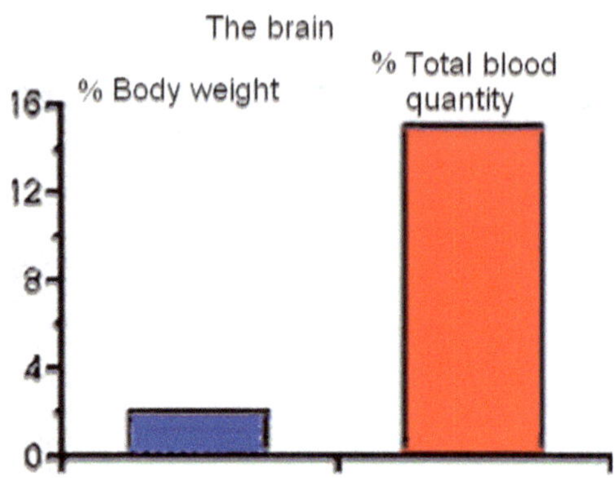

Figure 21.3 Blood circulation in the human brain.

The effect of meditation on the cardio-vascular function is now under scientific investigation and its favourable effects have been confirmed. Research continues encouraged by Dalai Lama himself.[6]

[5] Philip Rawson, *Tantra, The Indian Cult of Ecstasy*, Thames and Hudson, London, 1973 (Nepal, ca. 1700 CE).

[6] Leader of the ruling class Dge-lugs-pa (of the Yellow Hat) of Tibetan Buddhism and religious as well as political leader of Tibet until 1959, when independence of Tibet was abol-

21.2 The Homeric Meter

According to Francois Haas,[7] referring to reciting metrical Homeric poetry and its effects on human physiology, it is certain that all internal rhythms can be modified by external stimulations.

Recently, researchers from Austria, Germany and Switzerland, tested 20 healthy individuals, men and women, of average age 43, who were listening to repeated excerpts from a German translation of the Odyssey. Their heart and lungs were mechanically interconnected and their responses monitored.[8]

In the German edition of the Odyssey the complex rhythmic verse, the dactylic hexameter, is maintained. The latter consists of six parts, i.e., of a *long syllable* followed by a *long syllable* and a *short syllable* or of two *short syllables*. A detailed account follows.

While the patient was reading or listening to the verses, his respiration rate was decelerating, and heart and respiration rates were synchronizing more and more. These rhythms were fully de-synchronized when the patient stopped reading and started breathing normally again, returning to his/her every-day situation.

Concerning the above studies, i.e., the effect of Christian prayer by the use of rosaries or the utterance of Hindu or Buddhist mantras, it was found that respiration rate may drop to six per minute, assisting the heart to function more efficiently. In addition, the oxygen content in blood reaches saturation, which is an optimum condition.

In fact, scientists have been wondering whether rosaries have been so popular, because they make people feel better and more perceptive towards religious messages.

One of the scientists, Dirk Cysarz,[9] suggested that low respiration rate is associated with low blood pressure. Moreover, other investigations prove that low respiration rate leads to better lung function. It is emphasized that poetry must be recited properly in order to affect human body, e.g., mumbling is not

ished by Communist China. Present 14th Dalai Lama Bstan-'dzin-rgya-mtsho is Head of a Government in exile situated in Dharmsala, India, at the Himalayans. He is a Nobel Prize for Peace Laureate, thanks to his non-violent struggle for the independence of his country.

[7] Research Director, Cardiovascular Rehabilitation, Medical School, New York University, USA.

[8] Dirk Cysarz et al., Oscillations of heart rate and respiration synchronize during poetry recitation, *American Journal of Physiology – Heart and Circulatory Physiology*, online edition, **287**, 2, 2004.

[9] Department of Medical Theory and Complementary Medicine of Witten University, Herdecke, Germany.

helpful, while every syllable of the semi-verse must be carefully pronounced and every semi-verse followed by a relaxed respiration, to affect the heart rate.

Francois Haas adds that these phenomena are fully rational, without any trace of mystical dimension, since, in a living body, all organs are closely interconnected. He emphasizes the influence of marches accompanying rhythmic walking or rhythmic singing of South Asia natives, while rowing. In essence, all of these are techniques synchronizing respective functions.

21.3 The Dactylic Hexameter

The dactylic hexameter is a form of *metric* poetry or a rhythmic formation. Traditionally, it is related to classical poetry, mainly with the Homeric Epics, but also with Latin ones, such as Virgil's Aeneid.

Dactyl (δάκτυλος) is a system consisting of three syllables: The first one is *long,* the rest are *short.* Accordingly, the ideal verse of dactylic hexameter consists of six meters or feet, each one of which is dactylic. However, as a rule, the last foot of the verse is not pure dactylic, but rather a two-syllable *spondee* (σπονδεῖος) or *trochaic* (τροχαῖος), i.e., the syllable before the last is always long, while the last one is either long or short (such a syllable with optional accentuation is called an *anapaestic* – ἀνάπαιστος). In essence, it is difficult to arrange words according to this metre. So, poets often replace these dactyls with spondees, which are feet with two long syllables. Traditionally, the fifth foot of a verse is purely dactylic. About one out of twenty Homeric verses has a spondee at the fifth foot. Such a verse is called *spondaic*. For example, separation of verses into feet and semi-verses in the first seven verses of the Iliad is as follows:

μῆνιν ἄ/ειδε, θε/ά, || Πηληϊά/δεω 'Αχι/λῆος
οὐλομέ/νην, ἣ/μυρί' || 'Α/χαιοῖς / ἄλγέ, ἔθ/ηκεν,
πολλάς /δ' ἰφθί/μους || ψυ/χάς Ἄι/δι προί/αψεν
ἡρώ/ων, αὐ/τούς δὲ || ἑ/λώρια / τεῦχε κύ/νεσσιν
οἰω/νοῖσι τε /δαῖτα || Δι/ὸς δ' ἐτε/λείετο / βουλή,
ἐξ οὗ / δὴ τὰ π/ρῶτα || διαστή/την ἐρί/σαντε
'Ατρεΐ/δης τε ἄ/ναξ || ἀν/δρῶν καὶ / δῖος 'Α/χιλλεὺς

As a conclusion, based on the above findings and, in particular, on the last example, one can discover that reading the ancient Greek verse is an acoustic issue. People who can read dactylic hexameters according to grammar, cannot necessarily recite it loud and with a proper rhythm, therefore, the musicality of poetry is lost. Reading with sensitivity and satisfaction requires effort and training, while the initial rules that must be adhered to are few and simple. Certainly, the favourable effect on physical health is a worthy motive.

Appendix
The Forge – A Literary-Symbolic Approach

She was looking around totally lost. Her face was the face of pain. "A forge like any other", she thought. She watched more carefully: Bright armours hung from the walls, exquisite golden craters and cylices. "All of these cannot be made by human hand", she thought. Some strange tripods were moving around by themselves. They appeared to possess soul and mind. One of them moved silently towards her, and almost prompted her to sit on it. And then it was those weird golden girls. They looked like soulless metal dolls. But, at times, they grew alive and moved swiftly and accurately, obeying an eyeblick of the Master. At times, their expressionless faces grew tender, while rushing to assist him, when stumbling, betrayed by his crooked legs.

The Master himself was most weird. At first glance he looked like any other blacksmith. But the red hot metal in his hands was taking form and beauty, so fast that no eye could follow. He was the sovereign in the forge. Everything was obeying his eyes, everything was submitting to his will. Even fire was obeying him, and the bellows were producing air for the job by themselves. But even those hard, decisive eyes, at times grew dark, and tears were running down his scorched cheeks.

She noticed it and her heart ached. His matching with beautiful Charis proved an unfortunate mismarriage. It was not long, before Helios brought to him the ominous news. What if, by his uncomparable skill, he had eventually trapped and chained the unlawful lovers? The grief of betrayal cannot be easily erased from human hearts, but neither from the hearts of the immortals. The Master received her most warmly. He owed her his life. She had saved him, when his mother, after a violent arguing with his father, the Father of Gods and Men, had hurled him away from the summits of Olympus. She could not stand his lameness, so she sent him into the depths of the Great Sea, somewhere close to Lemnos island. And if it were not for the sweet Nereid and her sister Eurynome, many more trials would have burdened his soul. So he was pleased to see her entering. He wished that he could pay her back some day, and he even vowed to offer his life for her sake.

The Master knew the reason of her visit. He knew that the time had come for Achilles, her son, to return to battle. The Achaeans had paid his wrath very dearly. He had responded to their supplications by sending Patroclus, his dearest friend, to the battle, wearing his own armour. Just to trick the Trojans to believe that he was back. But fate had decided differently. Patroclus died from the hand of brave Hector, and Achilles' armour was lost. Now he was ready to rush to the battle, even with bare hands. Thetis, in despair, had come to ask the Master to fashion a new armour for her son. He knew it well. As he knew many more things.

Without lessening his efforts, his sharp eyes rested on her. The Nereid, the nymph, was a beauty, as always. Ethereal and crystal-fresh, like the element of nature that gave her birth. White, like the froth of the waves, mirroring sunset with her eyes and having moon's silver on her feet. Now he knew why Cronides, his allmighty father, was enchanted by her incomparable beauty and had tried to seduce her. He might have suceeded, if fear had not proved mightier than greatness. Because, when Prometheus, the Titan, told him of the ancient prophecy, that Thetis would bear a son greater and more glorious than his father, the father of gods and men stepped back. You see, even greatness has limits. Even among the greatest of gods, even concerning there own children. And for more certainty, when Thetis fell in love with Peleus and wanted to marry him, the gods decided that the son she would bear, would die in battle. Young age and love are never afraid of the future, even if the omens make it dark and abominable. Thetis ignored the prophecy. But long after her son's birth, she felt that it had begun to materialize: Achilles, even as a child, showed that he would be greater and more glorious than Peleus, his father, and his grand father Aeacos, to whose and Laomedon's hands Apollo and Poseidon had delivered the city of Troy. The city that, upon gods' orders, they had erected. Now the Great War had broken out, and her son was the greatest warrior among the Achaeans. Fear started growing in her, that, no matter what she had tried, fate had since long made her decisions. Maternal grief and despair were setting her in flames.

The Master stopped for a while his frantic hammering and looked at her thoughtfully.

– Are you sore, my Lady?

– I am scared, Master. I'm afraid that whatever I have devised to save him, was in vain. Fate is more powerful.
– Don't give up hope.
– I know, may be his time is not up yet. But look, gradually everything falls in place, like paving the road to the fulfillment of the prophecy.
– What prophecy?
– This war came so suddenly. Cursed Eris took her revenge. She was not invited to our wedding, and she let disaster loose. It was so nice to have Peleides hidden in Skyros, dressed as a girl, with the daughters of King Lycomedes. But that soothsayer, Calchas, prophesised that the war against Trojans could never be won without him. So they Achaeans started searching for him frantically. They would have never found him, if it were not for Odysseus, that sneaky fox. All he had to do was to put a weapon in front of the girls, and Peleides jumped to grasp it. From then on it was piece of cake to convince him to get to war. Even now, after all that has happened, he is ready to fight again.
– Is not your son invulnerable?
– Yes, when he was a baby, I bathed him in the waters of Styx. But a thought tortures me, that perhaps some part of him, may be the ankle I was holding him from, has remained out of the water.
– But the rest of his body is invulnerable.
– Indeed. But if the fate wants him dead, is enough. She can hit him even there. But I implore you, Master, hurry. My son is in the middle of the war. Unarmed.

The Master remained silent. He got to work again, hammering much faster now. His expression was grave, almost mournful. An invisible breath of life had filled the forge. Everything was following the rhythm of his hammer, each material, each tool, each blow of air, each flame, small or big, everything was vibrating as alive, everything was working harmoniously, to make the work perfect. The results soon appeared. One by one, the various parts of the shining armour were ready: the exquisite helmet, the shining breastplate, the fine greaves. Each one was emerging out of the hazy air, a divine creation of wisdom, strength and beauty.

Lastly, the lame technician fashioned the Shield. Hard bronze, tin and pure gold were wisely combined, to create an impenetrable wall, arresting swords and arrows and spears and scattering all impact in the wind, annihilating its momentum. Gold and silver were combined to create on the shield, as ornate as possible, designs greater and more magnificent than mortal eyes had ever seen.

Thetis saw it and remained speechless as lightning-stricken:

– So you knew!
– Yes, my Lady, I knew. Just as you knew.
– You are right. But I did not want to believe what I knew. And this is his end. Final and irrevocable.
– I finished the job, my Lady. There is nothing more to do. I had to give him what he needs to win. This is the beginning of the end of his enemies and Troy and of Achaeans' final triumph. In essence, all his own achievements. And the glory is his own too.
– And his end, the end of a son greater than any mother has ever given birth to. Alas, me the wretched, such a folly, such a flippancy, such senselessness before the prophecy! O love, how sweet, how abominable you are. How little you are worth before your own fruits!
– My Lady, you drive me crazy. I can't bear you suffering like this.
– And what can you do, good Master?
– I can do a lot for you, my Lady. Even give my own life.
– I know that, dear. But what's the use?
– One word from you and I know what to do.
– What are you saying? Is there any hope?
– I don't know if this is hope – I don't know how angry the Gods will get. But my skill can make the Achaeans win the war. Without your son's help.
– What are you saying?
– Watch, my Lady, these good maids. They are made of gold, but they know more and can do more than the strongest and bravest of mortal men. Look at them. They are sitting at my feet, obedient like sweet kittens. But, I assure you, my art can turn them to wild beasts, and can go to war in your son's stead. They shall be invulnerable, truly invulnerable, against all weapons that the mortals know or might ever invent.
– You must be joking, Master. But even if it is so, do you really believe that Peleides will tolerate a war, by which warriors do not look the enemy into the eyes? And, rightly you said, the gods will never allow something like this.
– I never said that.
– Then what are you trying to say?
– I know for sure, my lady, that this will happen some day. I know that the mortals will steal my art, without having the sense to command it. Neither your son's bravery and virtue. They will rather send armies of metal against their enemies, while they will be waiting for the victory,

in safe shelters. So they shall hope, that is, since fear will reside in their hearts for good, and will never feel safe neither peaceful again. Safe and peaceful can feel only those who dare to face the enemy on equal terms – hoping equally for life and death.

– But why should the gods ever consent to such a horror?

– They have already consented, my Lady. The war lives deeply in the hearts of the mortals, along with a mind that cannot be stopped even at infinity. Both Dionysus and the Titans have built within each one of them their own strongholds fighting each other incessantly. But so it is with gods. The war lives in them too, in their shining mansions, within Olympus himself and within the worlds above Olympus. Gods and mortals are so much alike that I can't tell who made whom. And if you ask how all this started, neither myself, nor most wise Pallas, not even Cronides, the gatherer of clouds, knows the answer. That tremendous and sublime answer!

– And you have needled that answer on the Shield, didn't you, Master?

– You got it, my Lady. Everything is on this Shield. Peace and war, peaceful life, dissension and pain, life and death, sun, moon and the stars. And all these within the great circle of the river ocean, enclosing diverse elements and showing that all is one: The supreme knowledge of Good and Evil and their absolute unity. A knowledge the mortals highly hate and despise.

– This is why the Shield is so scary. Watching it, you can feel the endlessness of infinity and shiver at the idea that, at any time, you may be dissolved in it.

– And, mind you, there is no enemy so brave, not to fear what you are now watching.

– Even I feel fear before this Shield.

– You, a goddess! Think of the mortals. Not even the fearless Myrmidons shall be able to look at it without remorse.

– But once you conceive its meaning, fear goes, is not so, Master?

– So it is, my Lady. But then there is nothing left to learn, neither there is need to live anymore. You are not even interested in living.

– And my son, Master? My son, what will he do?

– Your son was born with this knowledge, my Lady. He has lived all his life on a thread thinner than a hair. The thread separating life from death. So the gods decided, when your love for bright Peleus led you to defy your own race. Your son has tasted life to the utmost, being the great warrior he is. But now he wants all the same to taste death, since he was found to know things that not even gods know. Those who, involuntarily, have

> led him on the path of knowing. Don't grieve for your son, my Lady. Fearless and serene, he will receive the Shield, aware that here his world is ending. He will dress himself with the armour and peacefully cross the line separating him from the other worlds. But, eons from today, gods and mortals will be keeping his memory, search for him, interpret him, making him good or evil, depending on the way that each one will feel his great soul. But none of them will ever defy his glory. Do not grieve for your departing son, my Lady. If you want, you may grieve for all of us, who stay behind. Not for him.

The goddess, with expressionless face, grasped the shining armour silently and flew with swift wings towards the shores of Troy. Where he was waiting.

MIX
Papier aus verantwortungsvollen Quellen
Paper from responsible sources
FSC® C105338

If you have any concerns about our products,
you can contact us on
ProductSafety@springernature.com

In case Publisher is established outside the EU,
the EU authorized representative is:
Springer Nature Customer Service Center GmbH
Europaplatz 3, 69115 Heidelberg, Germany

Printed by Libri Plureos GmbH
in Hamburg, Germany